Springer Tracts in Advanced Robotics

Volume 50

Editors: Bruno Siciliano · Oussama Khatib · Frans Groen

Ron Alterovitz and Ken Goldberg

Motion Planning in Medicine: Optimization and Simulation Algorithms for Image-Guided Procedures

 Springer

Professor Bruno Siciliano, Dipartimento di Informatica e Sistemistica, Università di Napoli Federico II, Via Claudio 21, 80125 Napoli, Italy, E-mail: siciliano@unina.it

Professor Oussama Khatib, Robotics Laboratory, Department of Computer Science, Stanford University, Stanford, CA 94305-9010, USA, E-mail: khatib@cs.stanford.edu

Professor Frans Groen, Department of Computer Science, Universiteit van Amsterdam, Kruislaan 403, 1098 SJ Amsterdam, The Netherlands, E-mail: groen@science.uva.nl

Authors

Ron Alterovitz
UCSF Comprehensive Cancer Center
University of California, San Francisco
San Francisco, CA 94143-1708
USA

and

Department of Electrical Engineering and
Computer Sciences
University of California, Berkeley
Berkeley, CA 94720-1777
USA
E-Mail: ronalt@berkeley.edu

Professor Ken Goldberg
Department of Industrial Engineering and
Operations Research
Department of Electrical Engineering and
Computer Sciences
University of California, Berkeley
Berkeley, CA 94720-1777
USA
E-Mail: goldberg@berkeley.edu

ISBN 978-3-540-69257-7   e-ISBN 978-3-540-69259-1

DOI 10.1007/978-3-540-69259-1

Springer Tracts in Advanced Robotics   ISSN 1610-7438

Library of Congress Control Number: 2008928446

Typeset & Cover Design: Scientific Publishing Services Pvt. Ltd., Chennai, India.

Printed in acid-free paper

5 4 3 2 1 0

springer.com

STAR (Springer Tracts in Advanced Robotics) has been promoted under the auspices of EURON (European Robotics Research Network)

Dedicated to our families for steering us in the right direction

Foreword

Since the dawn of the new millennium, robotics has undergone a major transformation in scope and dimensions. This expansion has been brought about by the maturity of the field and the advances in its related technologies. From a predominantly industrial focus, robotics has been rapidly expanding into the challenges of the human world. The new generation of robots is expected to safely and dependably cohabit with humans in homes, workplaces, and communities, providing support in services, entertainment, education, healthcare, manufacturing, and assistance.

Beyond its impact on physical robots, the body of knowledge robotics has produced is revealing a much wider range of applications reaching across diverse research areas and scientific disciplines, such as: biomechanics, haptics, neurosciences, virtual prototyping, animation, surgery, and sensor networks among others. In return, the challenges of the new emerging areas are proving an abundant source of stimulation and insights for the field of robotics. It is indeed at the intersection of disciplines that the most striking advances happen.

The goal of the series of *Springer Tracts in Advanced Robotics* (*STAR*) is to bring, in a timely fashion, the latest advances and developments in robotics on the basis of their significance and quality. It is our hope that the wider dissemination of research developments will stimulate more exchanges and collaborations among the research community and contribute to further advancement of this rapidly growing field.

The monograph written by Ron Alterovitz and Ken Goldberg combines ideas from robotics, physically-based modeling, and operations research to develop new motion planning and optimization algorithms for image-guided medical procedures. A challenge clinicians commonly face is compensating for errors caused by soft tissue deformations that occur when imaging devices or surgical tools physically contact soft tissue. A number of methods are presented which can be applied to a variety of medical procedures, from biopsies to anaesthesia injections to radiation cancer treatment. They can also be extended to address problems outside the context of medical robotics, including nonholonomic motion planning for mobile robots in field or manufacturing environments.

As the first focused STAR volume in the growing research area of medical robotics, this title constitutes a fine addition to the series!

Naples, Italy,   Bruno Siciliano
April 2008   STAR Editor

Acknowledgments

Medical robotics brings together researchers and practitioners from a variety of backgrounds. It is with great pleasure that we acknowledge our collaborators on the research presented in this book.

We would especially like to thank Russ Taylor at Johns Hopkins University for introducing us to the challenges of needle insertion and tissue deformations, Allison Okamura for her key contributions to medical needle steering, and Jean Pouliot and I-Chow(Joe) Hsu from the UCSF Department of Radiation Oncology for introducing us to important problems in radiation treatment in medicine.

We would like to thank the entire needle steering team including Allison Okamura, Noah Cowan, Greg Chirikjian, Vinutha Kallem, Kyle Reed, and Sarthak Misra at Johns Hopkins University, Robert Webster at Vanderbilt University, and Gabor Fichtinger at Queens University.

We would like to thank our collaborators at UC Berkeley who have helped us with physically-based simulation and meshing, including James O'Brien, Jonathan Shewchuk, and Nuttapong Chentanez, as well as our collaborators in operations research and optimization, including and Alper Atamtürk, Andrew Lim, and Laurent El Ghaoui

We would like to thank our additional collaborators at UCSF in Radiation Oncology, Radiology, and Bioengineering, including John Kurhanewicz, Sue Noworolski, Chris J. Diederich, Etienne Lessard, Yongbok Kim, Richard Taschereau, Adam Cunha, and Frank Tendick.

We would like to thank our colleagues at LAAS-CNRS in Toulouse, France, regarding motion planning and SMR, including Nic Siméon, Juan Cortés, Georges Giralt, Jean-Paul Laummond, and Rachid Alami.

We would like to thank members of the international research community who have helped us in advancing the field, including Simon DiMaio (Intuitive Surgical), Septimiu (Tim) Salcudean (University of British Columbia), Cenk Cavusoglu (Case Western Reserve University), Michael Branicky (Case Western Reserve University), Frank van der Stappen (Utrecht University), Jaydev Desai (University of Maryland), Pierre Dupont (Boston University), Ruzena Bajcsy (University of California, Berkeley), Kimmen Sjölander (University of

California, Berkeley), Lydia Kavraki (Rice University), Henry Fuchs (University of North Carolina at Chapel Hill), Dinesh Manocha (University of North Carolina at Chapel Hill), Stéphane Cotin (INRIA, Lille, France), Cameron Riviere (Carnegie Mellon University), Xunlei Wu (CIMIT, Cambridge, MA), Tolga Goktekin (Pixar), and Jessica Crouch (Old Dominion University).

We would like to thank current and former students and post-docs in Ken Goldberg's Automation Sciences Lab with whom we have discussed research over the years, including Dezhen Song, Vincent Duindam, Jeremy Schiff, Jijie Xu, K. ÓGopalÓ Gopalakrishnan, Vijay Vasudevan, Michael Yu, Brian Sung Chul Choi, Ryan Chen, Ephrat Bitton, and Anthony Levandowski.

We would also like to thank Bruno Siciliano, Tom Ditzinger, and Heather King for their guidance on the publishing of this monograph in the *Springer Tracts in Advanced Robotics*.

Finally, we would like to thank the government agencies that have supported our research, including the National Institutes of Health (NIH R21 EB003452, R01 EB006435, F32 CA124138), the National Science Foundation (NSF Graduate Research Fellowship), and the Department of Defense (National Defense Science and Engineering Graduate Fellowship).

Berkeley, California,   Ron Alterovitz
April 2008   Ken Goldberg

Contents

1 Introduction 1
  1.1 Motion Planning for Image-Guided Medical Procedures 2
  1.2 Motion Planning Algorithms 4
    1.2.1 Motion Planning for Rigid Needles 4
    1.2.2 Motion Planning for Steerable Needles 5
    1.2.3 Motion Planning for Radiation Sources for Cancer
        Treatment 6
  1.3 Brachytherapy for Treating Prostate Cancer 6
  1.4 Contributions 8
  1.5 Overview 9

2 Physically-Based Simulation of Soft Tissue Deformations . . . 11
  2.1 Fundamentals of Continuum Mechanics 12
    2.1.1 Deformable Bodies 13
    2.1.2 The 1-D Case 13
    2.1.3 The 3-D Case 15
    2.1.4 The 2-D Case 18
  2.2 Simulating Soft Tissue Deformations 18
    2.2.1 Mass-Spring Method 18
    2.2.2 Finite Element Method 19
    2.2.3 Visualizing 2-D Simulations 23
  2.3 Conclusion 24

**3 Motion Planning in Deformable Soft Tissue with
   Applications to Needle Insertion** 27
  3.1 Sensorless Planning and Needle Insertion 29
  3.2 Problem Formulation 30
  3.3 Simulating Needle Insertion 31
    3.3.1 Background on Needle Insertion Modeling and
        Simulation 31
    3.3.2 Input Anatomy Model 32
    3.3.3 Simulation Output 33

        3.3.4   Simulating Needle Procedures 33
        3.3.5   Simulation Visualization 36
    3.4 Motion Planning for Needle Insertion 37
        3.4.1   Method Overview 37
        3.4.2   Planning Problem Formulation 37
        3.4.3   Planning Algorithm 38
    3.5 Application to Brachytherapy Cancer Treatment 39
        3.5.1   Simulation Implementation 40
        3.5.2   Sensorless Planner Results 41
    3.6 Conclusion and Open Problems 42

4 **Motion Planning in Deformable Soft Tissue with Obstacles**
   with Applications to Needle Steering 45
    4.1 Background on Needle Steering 45
    4.2 Simulating Needle Steering 47
        4.2.1   Soft Tissue Model 48
        4.2.2   Computing Soft Tissue Deformations 48
        4.2.3   Needle Insertion Model 48
        4.2.4   Simulating Cutting at the Needle Tip 49
        4.2.5   Simulating Friction Along the Needle Shaft 50
        4.2.6   Simulation Results 51
    4.3 Motion Planning for Needle Steering 52
        4.3.1   Problem Formulation 52
        4.3.2   Optimization Method 53
        4.3.3   Planner Results 54
    4.4 Conclusion and Open Problems 55

5 **Motion Planning for Curvature-Constrained Mobile**
   Robots with Applications to Needle Steering 57
        5.0.1   Uncertainty and Motion Planning 58
        5.0.2   Background on Nonholonomic Motion Planning and
                MDP's 59
        5.0.3   Overview of Motion Planning Method 61
    5.1 Problem Definition 62
    5.2 Motion Planning for Deterministic Needle Steering 64
        5.2.1   State Space Discretization 64
        5.2.2   Deterministic State Transitions 65
        5.2.3   Discretization Error 66
        5.2.4   Computing Deterministic Shortest Paths 66
    5.3 Motion Planning for Needle Steering under Uncertainty 66
        5.3.1   Modeling Motion Uncertainty 67
        5.3.2   Maximizing the Probability of Success Using Dynamic
                Programming 67
    5.4 Computational Results 69
    5.5 Conclusion and Open Problems 73

6   The Stochastic Motion Roadmap: A Sampling-Based Framework for Planning with Motion Uncertainty 75

    6.0.1   Related Work 78

    6.0.2   SMR Contributions 79

  6.1   Algorithm 80

    6.1.1   Input 80

    6.1.2   Building the Roadmap 81

    6.1.3   Solving a Query 82

    6.1.4   Computational Complexity 84

  6.2   SMR for Medical Needle Steering 84

    6.2.1   SMR Implementation 85

    6.2.2   Results 86

  6.3   Conclusion and Open Problems 89

7   Motion Planning for Radiation Sources during High-Dose-Rate Brachytherapy 91

  7.1   Introduction to HDR Brachytherapy and Dose Optimization ... 92

  7.2   Linear Programming for HDR Brachytherapy 93

    7.2.1   Patient Data Input 95

    7.2.2   Dose Calculation 95

    7.2.3   Clinical Criteria 96

    7.2.4   Linear Programming Formulation 97

  7.3   Application to Prostate Cancer Treatment 99

    7.3.1   Patient Data Sets 100

    7.3.2   Evaluation Metrics 100

    7.3.3   Results 101

  7.4   Discussion 104

  7.5   Conclusion and Open Problems 105

8   Conclusion 107

  8.1   Contributions 107

    8.1.1   Deformations 107

    8.1.2   Uncertainty 108

    8.1.3   Optimality 108

  8.2   Future Directions 108

    8.2.1   Realistic Simulation of Image-Guided Medical Procedures 108

    8.2.2   Planning Algorithms for Image-Guided Medical Procedures 111

    8.2.3   New Clinical Applications 113

  8.3   Conclusion 113

References 115

A   Target Localization Using Deformable Image Registration 129
  A.1 Introduction to Deformable Image Registration............... 129
  A.2 Deformable Registration with Model Parameter Estimation 132
    A.2.1 Method Input.. 133
    A.2.2 Finite Element Model of Soft Tissue Deformation 133
    A.2.3 Quality Metric 134
    A.2.4 Optimization of Uncertain Parameters 134
    A.2.5 Visualizing Registration Output 135
  A.3 Application to Prostate Cancer Treatment 135
    A.3.1 Patient Image Acquisition 138
    A.3.2 Application of the Deformable Registration Method 139
    A.3.3 Warping the MRSI Grid.............................. 141
    A.3.4 Method Evaluation and Parameter Selection 142
    A.3.5 Results... 142
    A.3.6 Discussion 144
  A.4 Conclusion and Open Problems 147

Index 149

1 Introduction

Emerging advances in medical imaging are enabling clinicians to non-invasively examine anatomy and metabolic processes deep below the skin surface. From computed tomography capable of displaying the patient's 3-D anatomy with sub-millimeter resolution, to magnetic resonance spectroscopy imaging that can identify the location of metabolic compounds in tissue, the quantity and detail of patient-specific imaging data available to clinicians is rapidly increasing.

In parallel to these advances in medical imaging, new robotic tools are being introduced into clinical practice. These "robotic surgical assistants" have the potential to provide greater precision and accuracy compared to manually controlled surgical devices. A pioneer in this area has been the commercially successful da Vinci Surgical System, a robotic surgical assistant for laparoscopic surgery developed by Intuitive Surgical that has been installed in over 700 locations worldwide. In addition to the da Vinci system, numerous robotic systems are being developed commercially and in academia for specialized medical procedures from biopsies to retinal surgery to radiation dose delivery.

Fully integrating the wealth of digital information obtained from imaging with advances in robotic hardware has the potential to significantly improve patient care. To fully realize this potential, new computational tools are needed to help physicians transform the information obtained from medical images into actions for robotic surgical assistants to perform.

In this book, we focus on new motion planning algorithms for image-guided medical procedures. These motion planning algorithms must address key challenges that arise in medical applications, including deformations, uncertainty, and optimality. The motion planning algorithms utilize anatomical and clinical information extracted from medical images as well as physically-based models of surgical devices and soft tissues. The objective of motion planning in this context is to compute actions that will guide a surgical device around anatomical obstacles to reach a clinical target or achieve a clinical goal.

R. Alterovitz and K. Goldberg: Motion Planning in Medicine, STAR 50, pp. 1–10, 2008.
springerlink.com © Springer-Verlag Berlin Heidelberg 2008

1.1 Motion Planning for Image-Guided Medical Procedures

Motion planning for image-guided medical procedures introduces three key algorithmic challenges:

1. *Deformations*: When surgical devices such as needles contact soft tissue, the soft tissue may deform. Clinicians must compensate for these deformations to successfully guide a surgical device to a clinical target.
2. *Uncertainty*: The motion response of surgical devices to commanded actions cannot always be predicted with certainty. Errors can arise due to patient variability as well as limitations inherent in the surgical device (for example, a "rigid" needle flexing due to contact with tissue). Clinicians must consider uncertainty to successfully guide surgical devices to clinical targets with a high probability of success.
3. *Optimality*: When multiple motion plans are feasible, how do we compute the best option? Almost all motion planning problems for medical procedures involve optimization to achieve the clinical goal as best as possible while minimizing tissue damage and other negative side effects.

To address these challenges, the computational tools introduced in this book combine biomedical imaging, physically-based simulation, and new geometric and optimization-based planning algorithms. This approach takes advantage of advances in robotics algorithms, finite element modeling, and operations research. Increases in computer processing speed are enabling the integration of results from these disparate fields in a novel fashion, allowing the creation and implementation of fast and effective motion planning algorithms for image-guided medical procedures.

We illustrate our proposed information flow for computer-assisted image-guided medical procedures in figure 1.1. Between the traditional image acquisition and treatment phases, a computational phase incorporates new algorithms to plan and optimize the procedure. Developing these computational tools is the focus of this book.

On the left of figure 1.1 is the acquisition of patient-specific information. Medical imaging serves as a key source of patient-specific information since images encode both anatomical structures as well as clinical targets. For example, molecular-scale imaging techniques such as Magnetic Resonance Spectroscopy Imaging (MRSI) enable physicians to non-invasively and precisely identify the location of anomalies such as cancerous lesions. Computed Tomography (CT) can display the patient's 3-D anatomy with a resolution as fine as 0.5 mm. Ultrasound imaging can display moving tissues in real-time. And X-ray fluoroscopy can image and localize surgical devices such as needles inside human tissue. In addition to images, additional patient-specific input must be specified by the physician. This includes clinical criteria, such as dose requirements for tissues during radiation cancer treatment.

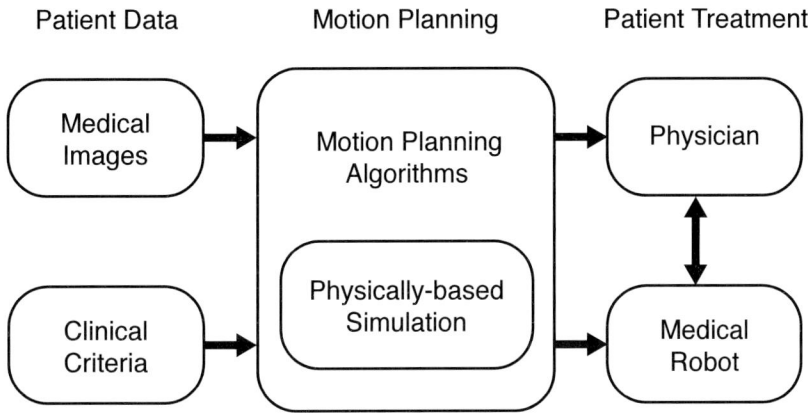

Patient Data   Motion Planning   Patient Treatment

Fig. 1.1. Information flow for computer-assisted image-guided medical procedures. In this book, we focus on the middle phase, computational tools for motion planning.

On the right of figure 1.1 is the patient treatment. Patient treatment can be performed by the physician directly by manually controlling surgical devices or with the help of robotic surgical assistants. In both cases, the surgical tools are subject to their own set of kinematic constraints, which must be considered during procedure planning and optimization.

In the middle of figure 1.1 is motion planning, the process of using the patient-specific information on the left to determine actions to be performed by the surgical devices on the right. As discussed above, the key challenges for motion planning in these contexts are deformations, uncertainty, and optimality.

To properly consider the effects of tissue deformations, the motion planning algorithm can utilize a physically-based simulation of the interaction of the surgical device with soft tissue. In this book, we introduce computational models and simulations of soft tissue deformation using finite element methods (FEM). Although FEM is generally used for off-line simulation of stiff solid materials, we harness the power of modern computers and new algorithmic techniques to perform real-time FEM simulation of soft tissues (as introduced in chapter 2). We demonstrate how these simulations can be used as components of higher-level planning algorithms, such as motion planning algorithms to compensate for errors caused by tissue deformations during needle insertion (chapters 3 and 4) as well as image registration algorithms to map targets across images (appendix A).

To properly consider the effects of uncertainty, we introduce motion planning algorithms that compute actions to maximize the probability that the surgical device will avoid obstacles and successfully reach the clinical target. These algorithms explicitly consider uncertainty in the motion response of a surgical tool due to patient variability and the complexity of tool/tissue interaction (chapters 5 and 6). In these cases, the information flow in figure 1.1 becomes a closed loop; medical images are obtained during patient treatment and the actions determined by the motion plan are automatically updated and passed to the physician and/or robotic surgical assistant.

Throughout this book, we focus on optimization-based motion planning. When multiple plans are feasible, criteria such as minimizing side effects or tissue damage can be used to determine the optimal plan. In cases such as radiation treatment where it is impossible to precisely satisfy all the physician specified dose prescriptions, optimization can be used to compute a plan that satisfies the dose prescriptions as best as possible.

In this monograph, we use one application as an illustrative example: prostate brachytherapy, a medical procedure for treating prostate cancer in which physicians use needles to place radioactive seeds in close proximity to cancerous cells. However, the approach outlined in figure 1.1 applies to a broad spectrum of medical procedures where disease is localized, from biopsies to anesthesia injections to chemical and thermal cancer treatments.

1.2   Motion Planning Algorithms

We focus on three motion planning problems that arise during image-guided medical procedures: motion planning for rigid needles, motion planning for steerable needles, and motion planning for radiation sources for cancer treatment. Each of these problems introduces new computational challenges and is subject to unique planning and optimization constraints imposed by the physician's treatment requirements, the patient's anatomy, and the physical limitations of medical equipment and devices. We present motion planning algorithms for each of these general problems and then customize the solution to the specific application of prostate brachytherapy.

1.2.1   Motion Planning for Rigid Needles

With the increasing use of minimally invasive image-guided medical interventions, needle insertion is becoming ubiquitous in modern medical procedures, from biopsies to anesthesia injections to cancer treatments such as cryotherapy and brachytherapy. Accurately guiding a needle to a specific target inside soft tissue is crucial for the success of these procedures. However, significant errors are common in current practice. A key source of these errors is the deformation of soft tissue caused by the needle during insertion.

We introduce a motion planning algorithm to compensate for errors caused by soft tissue deformation during needle insertion. The planner combines a software simulation of tissue deformations that occur during needle insertion with an optimization algorithm to determine the needle entry location and insertion depth that minimizes errors. We apply the planner to an example from prostate brachytherapy, where the success of the procedure depends on the accurate placement of radioactive seeds within the prostate gland [62, 178] and ignoring these deformations can result in misplaced seeds [178, 181].

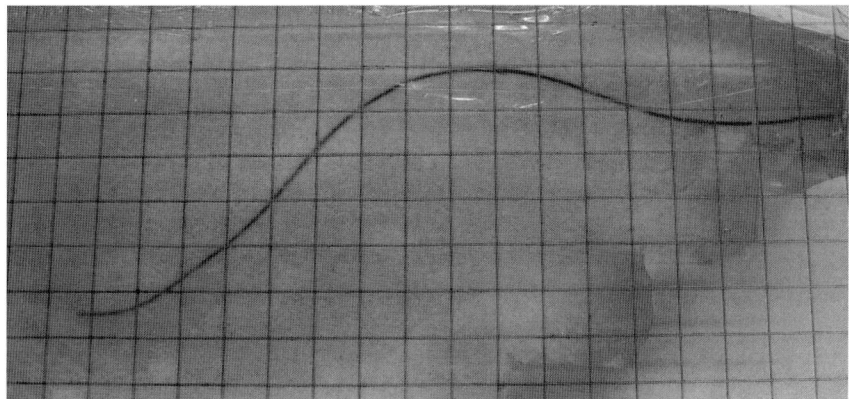

Fig. 1.2. A prototype steerable needle is inserted into an artificial tissue gel. The lines in the overlaid grid are separated by 1cm. Steerable needles are capable of following curved paths around obstacles to reach targets inaccessible to traditional straight, rigid needles.

1.2.2   Motion Planning for Steerable Needles

Steerable needles are a new class of highly flexible bevel-tip medical needles that can be *steered* around obstacles to targets in soft tissue previously inaccessible to rigid needles [211]. We demonstrate the motion of a steerable needle in figure 1.2. To fully harness the potential of these new needles, motion planning algorithms must consider these needles' nonholonomic constraints and the uncertainty in their motion through soft tissue. Needle steering can be viewed as a type of nonholonomic motion planning for a car-like mobile robot.

We introduce a motion planning algorithm for steerable needles that finds obstacle-free paths to the target while compensating for errors caused by soft tissue deformation. As in the rigid needle case, we develop an optimization-based motion planner that uses a software simulation of tissue deformations. For steerable needles, we generalize the objective function of the optimization algorithm to minimize costs due to insertion length and obstacle collision, as well as placement errors caused by tissue deformations.

We also consider motion uncertainty for steerable needles due to patient variability and the complexity of needle/tissue interaction. To address uncertainty, we introduce methods that compute actions to maximize the probability that the steerable needle will avoid obstacles and successfully reach the target. We first introduce a method specialized to devices that follow constant curvature paths. We then generalize this method and introduce the Stochastic Motion Roadmap (SMR), a new motion planning framework that explicitly considers motion uncertainty during planning by combining motion sampling with Markov Decision Processes and Dynamic Programming. We apply the SMR framework to needle steering and show that accounting for needle motion uncertainty during planning can significantly increase the probability of reaching targets without colliding with obstacles.

1.2.3   Motion Planning for Radiation Sources for Cancer Treatment

Using medical images of patient anatomy and estimates of tumor location, physicians prescribe radiation dose requirements for cancerous tumors and surrounding tissues. The dose can be delivered by moving a radiation source through needles implanted in or near the cancerous tissues. Since dose increases linearly with time, controlling the speed at which the source moves through a particular region of tissue determines the dose delivered to that tissue. The motion planning challenge is to determine how to move the source through needles to achieve the physician-specified dose requirements as best as possible.

We introduce an optimized-based motion planning method for radiation sources to optimize dose delivered to the patient. Our method, based on linear programming (LP), is fast and exact and computes radioactive source locations and dwell times to maximize the satisfaction of physician specified dose requirements. The method uses the objective and clinical criteria framework of Inverse Planning by Simulated Annealing (IPSA), an approach developed at the University of California, San Francisco (UCSF) that has been used in the treatment of over a thousand patients. Unlike previous methods used for dose optimization, the LP method guarantees a mathematically optimal solution.

1.3   Brachytherapy for Treating Prostate Cancer

Prostate cancer kills over 30,000 Americans each year [163]. It is the second leading cause of cancer death for men in the United States (after lung cancer). One in six American men will be diagnosed with prostate cancer during their lifetime, and someone will die from it approximately every 18 minutes [113].

The prostate is a gland roughly the size of a walnut. It is shaped like a pyramid, with average transverse × anteroposterior × craniocaudal dimensions of 4 cm × 3 cm × 3 cm [48, 121]. The prostate is located inferior to the bladder and anterior to the rectum and surrounds the urethra, as shown in figure 1.3.

Prostate cancer is often treated with brachytherapy, a minimally invasive medical procedure in which physicians place radioactive seeds in close proximity to cancerous tumors. Unlike other radiation treatments such as external beams, the radioactive source in brachytherapy is placed a short distance from the tumors ("brachy" means short in Greek). Because of this, brachytherapy can effectively be used to deliver a high dose to the cancerous tumor and a low dose to surrounding healthy tissue.

In brachytherapy, the radiation sources are radioactive seeds, approximately 4 mm long and 0.8 mm in diameter. The seeds are guided to their destination using hollow medical needles. Using medical images, the physician prescribes radiation doses for the prostate and surrounding tissues. The radioactive dose delivered by the seeds should "conform" to the physician specified prescriptions over the patient anatomy. Past studies indicate that improving radiation dose conformality improves patient health and reduces complication rates [110, 127, 203].

Two variants of prostate brachytherapy are commonly used in clinical practice: Low Dose Rate (LDR) and High Dose Rate (HDR). Because the response

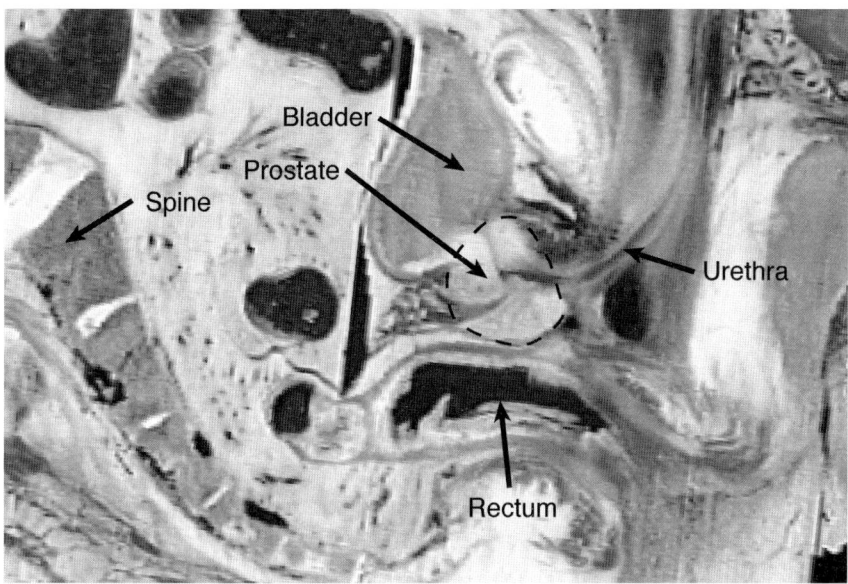

Fig. 1.3. The prostate is shown in the sagittal plane for a man laying on his back (the cranial direction is to the left). The urethra passes through the prostate and connects to the bladder adjacent to the prostate. This image was obtained from the National Library of Medicine's Visible Human project [201].

of cancer cells to radiation depends on dose rate, which variant of the procedure is used for a particular patient depends on many factors, including the location and stage of the prostate cancer and other medical considerations [160].

In low dose rate (LDR) brachytherapy, typically reffered to as *permanent seed brachytherapy* or *permanent prostate implant (PPI) brachytherapy*, physicians use needles to permanently implant low dose rate radioactive seeds inside the prostate, which will irradiate the prostate and surrounding tissue over several months. Prior to implantation, a CT or MR image is obtained of the patient anatomy and a dosimetric plan is prepared that specifies seed locations inside the prostate that best satisfy physician specified dose requirements, such as a minimum peripheral dose coverage, a uniform dose distribution inside the prostate gland, protection of the urethra, and a dose boost to the tumor. Approximately 100 seeds and biodegradable spacers are loaded into 20 to 25 needles. The physician inserts each needle transperineally into the patient, who is lying on his back as shown in figure 1.4. Seeds and spacers are pushed out of the needle when the depth of the needle specified by the dosimetric plan is reached. Achieving the desired seed placement is left to the physician, who must take into account factors such as needle bending and tissue deformations during the implant process [62, 178, 181].

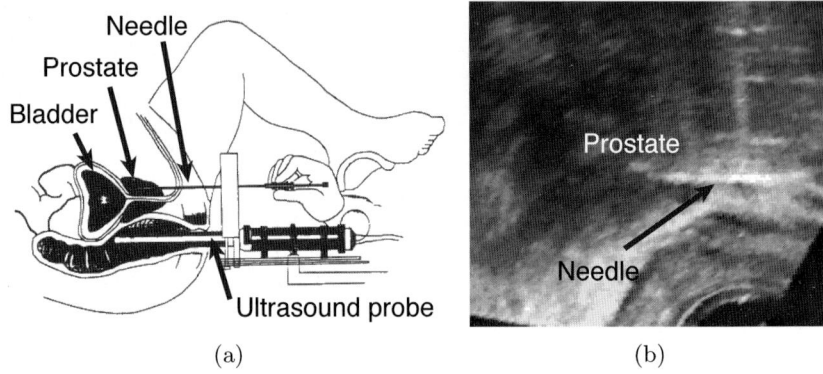

Fig. 1.4. During permanent seed brachytherapy, needles carrying radioactive seeds are inserted transperineally into the patient (a), who is lying on his back [178]. The quality of intra-operative transrectal ultrasound images is very poor (b), making it difficult to monitor deformations of soft tissues such as the prostate.

In *High Dose Rate (HDR) brachytherapy*, physicians insert catheters into the prostate through the perineum under ultrasound guidance in the operating room. Approximately 18 catheters are required to offer sufficient dwell time positions to cover the entire prostate. A CT or MR image is then obtained of the patient anatomy and the physician prescribes dose requirements for each point in the prostate and surrounding tissues. A plan is then created which specifies seed dwell positions and dwell times that best satisfy the physician prescribed dose requirements. To execute the plan, the catheters are attached to a robot (such as a Nucletron MicroSelectron High-Dose-Rate Remote Afterloader) for treatment delivery. The robot moves a single radioactive source inside each catheter to each dwell position for the pre-computed dwell times [146]. This procedure may be repeated before the catheters are removed.

1.4   Contributions

The primary contributions presented in this monograph are new algorithms that computationally plan and optimize image-guided medical procedures based on medical images and physician-specified clinical criteria. With the exception of motion planning for steerable needles, which have not yet been approved by the U.S. Food and Drug Administration (FDA) for human trials, all algorithms developed in this book have been tested with data from human patients based on clinical medical images. In pursuit of this research, we:

- Identified and implemented appropriate models and algorithms to interactively simulate soft tissue deformations due to forces applied during surgical and interventional medical procedures. The simulation software integrates methods from real-time physically-based simulation in computer graphics and classical finite element methods.

- Developed a 2-D simulation of medical needle insertion. The simulation estimates tissue deformations using a finite element method and real-time mesh maintenance.

- Designed and implemented a 2-D deformable image registration method that explicitly considers tissue deformations when mapping targets between images. Results using prostate medical images indicate a statistically significant improvement in registration accuracy compared to previous methods.

- Designed and implemented a motion planning algorithm for traditional rigid needle insertion procedures to correct for errors caused by predicted soft tissue deformations. The method combines a finite element model of needle insertion in soft tissue with numeric optimization.

- Designed and implemented a motion planning algorithm for steerable needles to reach a target while avoiding obstacles and correcting for errors caused by predicted soft tissue deformations. The method combines a finite element model of needle steering in soft tissue with numeric optimization.

- Designed and implemented a motion planning algorithm that explicitly considers uncertainty in motion for nonholonomic mobile robots subject to a constant turning radius, and applied the planner to steerable needles. The algorithm combines geometric planning with Markov Decision Processes and Dynamic Programming. Results indicate that traditional shortest paths do not maximize the probability of successfully reaching the target when the needle's response to controls is not known with certainty.

- Developed the Stochastic Motion Roadmap (SMR), a general sampling-based motion planning framework that explicitly considers motion uncertainty to maximize the probability of success. We apply the framework to steerable needles, enabling a more complex representation of motion uncertainty.

- Formulated the HDR brachytherapy dose optimization problem as a linear program, enabling the fast computation of mathematically optimal solutions. The linear program uses the objective and clinical criteria framework developed at the University of California, San Francisco (UCSF) and maximizes satisfaction of physician specified dose prescriptions.

- Used the optimal HDR brachytherapy solutions obtained by our linear program as a baseline to statistically validate the optimization performance of current clinical software (based on the probabilistic optimization method of simulated annealing) that has been used in the treatment of over a thousand cancer patients internationally.

1.5   Overview

In chapters 2 through 7, we introduce new motion planning algorithms for computer-assisted image-guided medical procedures, as illustrated in the middle phase in figure 1.1. In chapter 2, we present biomechanical models of soft tissue and develop a physically-based simulation of soft tissue deformation based on a finite element method. This simulation tool will serve as a building-block for the planning and optimization algorithms presented in subsequent chapters. In addition, we illustrate in appendix A how to use this simulation as a component in an image registration algorithm to automatically map targets identified in one image to their corresponding locations in another image in which the soft tissues have deformed.

In chapter 3, we develop a motion planning algorithm that combines a simulation of soft tissue with numerical optimization. We apply the method to traditional needle insertion procedures to correct for tissue deformation caused by forces exerted by the needle. We extend this planner and simulation in chapter 4 to consider the curved paths and obstacle avoidance made possible by steerable needles.

In chapter 5, we consider motion planning for steerable needles in the presence of uncertainty by using a constant-curvature model of needle steering. In chapter 6, we generalize this method and introduce the Stochastic Motion Roadmap (SMR), a new sampling-based framework that explicitly integrates a motion uncertainty model into the planning algorithm to maximize the probability of success. We apply the SMR framework to needle steering and illustrate the advantages of SMR compared to solving for traditional shortest paths that ignore motion uncertainty.

In chapter 7, we develop a motion planning algorithm for radiation sources that optimizes radioactive source locations and dwell times for high-dose-rate brachytherapy prostate cancer treatment. The method uses linear programming to optimally satisfy physician-specified dose requirements.

Finally, in chapter 8, we conclude and suggest future research directions where the combination of imaging data and physically-based simulation with planning and optimization algorithms have further potential to improve patient care.

2 Physically-Based Simulation of Soft Tissue Deformations

Physicians are increasingly performing surgical procedures using minimally invasive instruments that operate inside the body through narrow openings. This reduces disturbance to healthy tissue, minimizes risk of infection, and speeds recovery. However, these procedures are often challenging for physicians to visualize and perform due to reduced visual and tactile feedback compared with traditional open surgical procedures. Fast and accurate computer simulations of these procedures can facilitate physician training and assist in pre-operative planning and optimization.

Surgery simulation creates a virtual environment in which a physician can interact with organs and tissues that are simulated on a computer. Simulations are being developed for a wide array of medical procedures, including laparoscopic surgery [189], bronchoscopy [42], and endoscopic surgery [24]. Surgery simulation aims to complement the traditional apprenticeship model of physician training; physicians can train in a controlled environment that exposes them to both common and rare cases and can practice new techniques without risks to patient safety. Studies indicate that surgical skills learned using computational simulators directly improve operating room performance by significantly decreasing procedure time and reducing the number of medical errors [88, 184, 189]. In one videotaped study on gallbladder dissection, physicians trained using surgery simulation performed the task 29% faster and with six times fewer errors than traditional training [189].

In addition to training, surgery simulation can also be applied to medical procedure planning. With patient-specific imaging data and a sufficiently realistic simulation of a procedure, a planner can search the space of possible tissue/tool interaction sequences to identify a plan that is best suited to accomplish the clinical objectives. The ultimate goal is to provide a pre-operative plan, integrated with medical imaging, to the physician or robotic hardware that will perform that procedure [182, 183, 199].

Just as flight simulators give pilots an opportunity to learn and practice flying in a variety of visibility and weather conditions, surgery simulators aim to allow physicians to perform a procedure "virtually" on a computer to practice on difficult patient cases without risking patient safety. But whereas flight simulation

R. Alterovitz and K. Goldberg: Motion Planning in Medicine, STAR 50, pp. 11–25, 2008.
springerlink.com © Springer-Verlag Berlin Heidelberg 2008

requires models of airflow and rigid objects such as the plane, landforms, and buildings, the key challenge in surgery simulation is simulating deformable tissue interactively. Accurately simulating and displaying tissue deformations and tool-tissue interactions in real-time poses a computationally challenging problem and is the topic of much current research.

In this chapter, we combine methods from classical finite element methods with recent approaches from computer graphics to create a real-time interactive simulation of soft tissues. The simulation achieves sufficient accuracy to warrant further investigation for clinical applications. We use the simulation of deformable soft tissues as a building block in chapter 3 to simulate and plan needle insertion procedures and in chapter 4 to simulate and plan needle steering. In both of these cases, accurately guiding a needle to a specific target inside the human body is crucial for the success of the procedure. However, significant errors are common in current practice due to soft tissue deformations.

In this chapter, we first provide background on simulation of deformable objects before presenting our simulation of soft tissues. In section 2.1, we provide an introduction to continuum mechanics, a mathematical framework that has successfully been used to characterize living tissues and their deformations under applied forces. We then discuss research on soft tissue simulation in section 2.2. Next, we present a finite element method for interactively simulating 2-D tissue deformations, including real-time visualization using texture-mapping. This simulation will serve as a building block for the simulators, planners, and registration tools in chapters 3 and 4 and appendix A.

2.1   Fundamentals of Continuum Mechanics

Continuum mechanics aims to describe the effect of external forces or disturbances on the global behavior of solids, liquids, and gases. The theory behind continuum mechanics was originally developed in the early nineteenth century by Claude-Louis Navier, Siméon Denis Poisson, and George Green [213]. It has since been successfully applied in a broad variety of domains, from airplane and bridge design to animating feature films. Since living tissue is composed of discrete cells, which in turn are composed of molecules and atoms, living tissue, like other materials, is not purely continuous. However, a large class of living tissues has been successfully characterized using the methods of continuum mechanics [44, 46, 86, 126].

In continuum mechanics, we assume that field quantities such as the densities of mass, velocity, and energy are continuous over time and space inside the material [87]. We will consider a *deformable body*, a continuous material within a closed surface. We can use continuum mechanics to study how such a deformable body behaves when it is subjected to external influences such as forces or temperature changes.

In this section, will describe the fundamentals of continuum mechanics. We start by formally defining a deformable body. We then introduce the basic concepts of continuum mechanics using a simple 1-D example, and then generalize to 3-D and 2-D deformable bodies. We will use the framework of continuum mechanics to compare and analyze methods for simulating soft tissue deformations in section 2.2.

2.1.1   Deformable Bodies

We consider a deformable body B, a continuous material within a closed surface, which is a subset of the space $\Re^n$ where $n \in \{1, 2, 3\}$. The deformable body is composed of a set of material points $\mathbf{p} \in B$. The initial geometry of the deformable body is its *reference state*.

External forces applied to a deformable body B may cause material points $\mathbf{p} \in B$ to move, resulting in a deformed body B'. B' specifies the geometry of the *deformed state* of the deformable body.

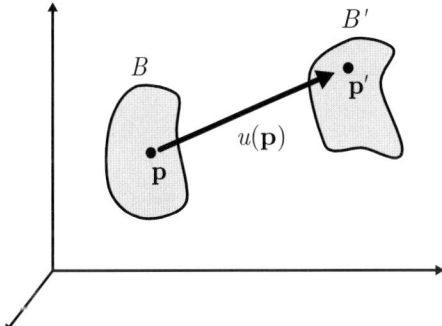

Fig. 2.1. A deformable body B in its reference state is deformed to B' by displacement field u

The *displacement* of a material point is its position change from B to B'. We define the transformation of a body B to a deformed state B' by a *displacement field*, which specifies the displacement for each material point in B, as shown in figure 2.1. Each point $\mathbf{p} \in B$ is transformed to a new point $\mathbf{p}' \in B'$ such that $\mathbf{p}' = \mathbf{p} + u(\mathbf{p})$, where $u(\mathbf{p})$ specifies the displacement field for each $\mathbf{p} \in B$. In continuum mechanics, we assume the mapping u that transforms B to B' is single valued, continuous, and has a unique inverse [87].

In dynamic simulations of deformable bodies, the displacement field $u(\mathbf{p})$ is a function of time. At all times, the displacement field is defined with respect to the original reference state. We define coordinates for points in the reference state in the *material coordinate frame*. The coordinates of displaced points during the simulation are defined in the *world coordinate frame*.

2.1.2   The 1-D Case

We introduce fundamental concepts from continuum mechanics with a simple 1-D example, a bar constrained along the x-axis. As illustrated in figure 2.2(a),

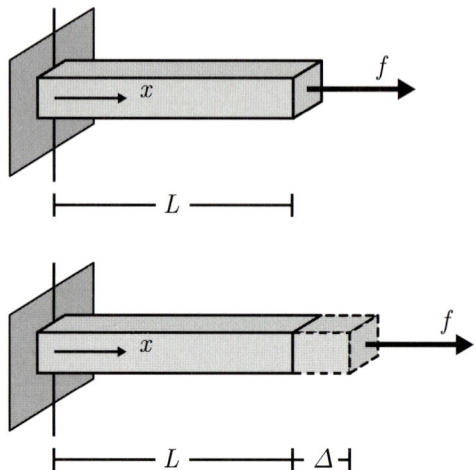

Fig. 2.2. A 1-D bar of length L in the reference state (top). Due to force f, the bar extends to length $L + \Delta$ in the deformed state (bottom).

the bar is fixed (i.e., attached to a wall) on the left at $x = 0$. At rest, the bar has length L. When a force f is applied to the bar along its axis, the bar will deform. If f points along the positive x-axis, the bar extends by a distance Δ, as shown in figure 2.2(b).

Stress is a measurement of force intensity: the total force acting on a surface divided by the area of that surface. Unlike force, which is a global quantity acting on the entire deformable body, stress is defined pointwise. In the 1-D bar example, stress σ at point x is defined by

$$\sigma(x) = \frac{f}{A},$$

where A is the (infinitesimal) cross-sectional area of the bar. Stress has units force per unit area. In the SI measurement system, stress has units of pascals (Pa), where the pascal is a derived unit equivalent to one newton per square meter.

Stress may result in a deformation of the deformable body. *Deformation* is measured as relative displacement, or

$$u(x + \Delta x) - u(x).$$

Strain is a measurement of relative deformation at a point. For a point x on a 1-D bar, strain ϵ is the ratio between the change in length of a segment (of infinitesimal length about x) and the original length of the segment:

$$\epsilon(x) = \frac{u(x + \Delta x) - u(x)}{\Delta x}.$$

In the limit as $\Delta x \to 0$,

$$\epsilon(x) = \frac{du}{dx}.$$

Strain has units of length per unit length, which is effectively unitless.

The relationship between stress and strain depends on the underlying material of the deformable body. We mathematically represent this relationship using a *constitutive relation*. In general, the constitutive relation is determined through physical experiments [86]. When the relation between stress and strain is linear, the material is linearly elastic and

$$\sigma = E\epsilon,$$

where E is the Young's modulus, a property of the material [87]. This relation is often referred to as Hooke's Law.

Given the geometry of the 1-D deformable body, the constitutive relation of the material, and the external applied forces, we can compute the resulting displacement field for the deformable body. We accomplish this by defining the stress resulting from the applied forces, computing the strain by plugging the stress into the constitutive relation, and integrating over the volume of the deformable body. For the 1-D bar example where the bar is composed of a linearly elastic material, the elongation Δ of the bar is computed by integrating strain over the length of the bar:

$$\Delta = \int_0^L \frac{du}{dx} dx = \int_0^L \epsilon(x) dx = \int_0^L \frac{\sigma(x)}{E} dx = \int_0^L \frac{f}{AE} dx = f \frac{L}{AE}. \qquad (2.1)$$

The quantity

$$k = \frac{f}{\Delta} = \frac{AE}{L}$$

is the *stiffness* of the bar. In 1-D, stiffness is a function of the Young's modulus and the geometry of the bar. For a linearly elastic material,

$$f = k\Delta.$$

Stiffness represents the amount of force required to achieve a unit displacement.

2.1.3   The 3-D Case

In 3-D, the relationships between stress, strain, and the constitutive relation are equivalent to the 1-D case. However, stress and strain are represented by tensors with 6 degrees of freedom rather than by scalars. Detailed derivations of the formulas for these tensors are available in standard continuum mechanics texts [87]. Here we focus on the fundamentals that will be applicable to soft tissue simulation in section 2.2.

To define stress at a point in 3-D, we consider an infinitesimal cube centered about the point. We illustrate the 9 components of stress at a point in 3-D in figure 2.3. Stress σ is defined by a 3×3 tensor

$$\sigma = \begin{bmatrix} \sigma_{11} & \sigma_{12} & \sigma_{13} \\ \sigma_{21} & \sigma_{22} & \sigma_{23} \\ \sigma_{31} & \sigma_{32} & \sigma_{33} \end{bmatrix},$$

where indices 1, 2, and 3 correspond to the x, y, and z axes, respectively. The elements along the diagonal are the *normal* stress components, while the off-diagonal elements are the *shear* stress components. The stress tensor is symmetric, resulting in 6 unique components [87].

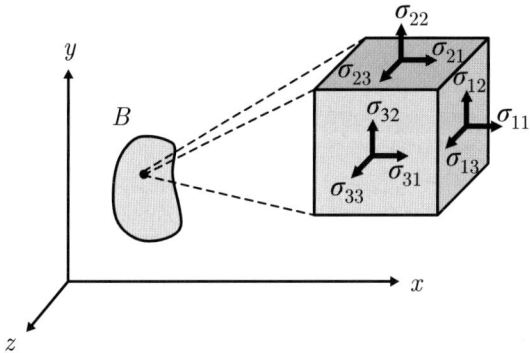

Fig. 2.3. The components of the stress tensor σ in 3-D for an infinitesimal cube from a deformable body B

Strain in 3-D is also defined by a 3×3 symmetric matrix with 6 unique components [87]:

$$\epsilon = \begin{bmatrix} \epsilon_{11} & \epsilon_{12} & \epsilon_{13} \\ \epsilon_{21} & \epsilon_{22} & \epsilon_{23} \\ \epsilon_{31} & \epsilon_{32} & \epsilon_{33} \end{bmatrix}.$$

Given the point-wise displacement field u for the deformable body, strain in 3-D can be computed similarly to 1-D by measuring the change in length of an infinitesimal segment. For a point $\mathbf{p}$ with displacement $\mathbf{u} = u(\mathbf{p})$, strain is a quadratic function:

$$\epsilon_{11} = \frac{\partial u_1}{\partial p_1} + \frac{1}{2} \left[\left(\frac{\partial u_1}{\partial p_1} \right)^2 + \left(\frac{\partial u_2}{\partial p_1} \right)^2 + \left(\frac{\partial u_3}{\partial p_1} \right)^2 \right] \tag{2.2}$$

and

$$\epsilon_{12} = \frac{\partial u_1}{\partial p_2} + \frac{\partial u_2}{\partial p_1} + \left[\frac{\partial u_1}{\partial p_1} \frac{\partial u_1}{\partial p_2} + \frac{\partial u_2}{\partial p_1} \frac{\partial u_2}{\partial p_2} + \frac{\partial u_3}{\partial p_1} \frac{\partial u_3}{\partial p_2} \right] \tag{2.3}$$

and similarly for the other strain components. Although quadratic strain is necessary to accurately model large rotations [226], higher order strain terms are often dropped to define the simpler "geometrically linear" strain:

$$\epsilon_{11} = \frac{\partial u_1}{\partial p_1} \tag{2.4}$$

$$\epsilon_{12} = \frac{\partial u_1}{\partial p_2} + \frac{\partial u_2}{\partial p_1} \tag{2.5}$$

which is applicable to smaller deformations without large rotations.

Once an appropriate representation of strain is selected, we can relate stress to strain using a constitutive relation that is appropriate for the material composing the deformable body. For a linearly elastic material,

$$\sigma_{ij} = \sum_{k=1}^{3} \sum_{l=1}^{3} C_{ijkl} \epsilon_{kl}$$

where $\mathbf{C}$ is a tensor of 81 elastic coefficients. For isotropic materials, tensor $\mathbf{C}$ can be derived from only two independent values: the Young's modulus E and Poisson's ratio ν. The constitutive relation in 3-D for isotropic linearly elastic materials is:

$$\sigma_{ij} = \sum_{k=1}^{3} \left(\frac{E\nu}{(1 - 2\nu)(1 + \nu)} \right) \epsilon_{kk} \delta_{ij} + 2 \left(\frac{E}{2(1 + \nu)} \right) \epsilon_{ij}$$

where δ_{ij} is the Kronecker delta function:

$$\delta_{ij} = \begin{cases} 1 : i = j \\ 0 : i \neq j \end{cases}.$$

As in the 1-D case, the Young's modulus is a measure of the stiffness of a material. Poisson's ratio is a measure of compressibility; when an object is stretched, Poisson's ratio quantifies the object's tendency to become thinner. In nonlinear methods such as Kelvin-Voigt, the tensor $\mathbf{C}$ can be functions of strain ϵ and strain rate $\dot{\epsilon}$ [87, 227].

Given the relationship between stress and strain, in 1-D we were able in equation 2.1 to obtain an analytic closed-form expression for the displacement of a point on the bar due to an external force by integrating over the volume of the bar. In higher dimensions, obtaining an analytic expression relating displacement to external forces is not possible, except for a small number of geometrically simple problems. To compute displacements for geometrically complicated deformable bodies, numerous methods have been developed to numerically compute approximate solutions, including mass-spring methods [49], boundary element methods [112], finite difference methods [49], and finite element methods [227]. In section 2.2, we focus on methods applicable to soft tissue deformations.

2.1.4   The 2-D Case

For certain problems, the stress and strain tensors defined for 3-D analysis can be simplified for 2-D analysis. One common approximation is *plane strain*, in which we assume $\epsilon_{33} = \epsilon_{23} = \epsilon_{31} = 0$. This assumption is valid when the object does not substantially displace or deform in the z-direction, which commonly occurs when the z-direction dimension of the body is large or restrained from motion. In 2-D surgery simulation, plane strain is appropriate if the tissue does not deform normal to the selected imaging plane.

2.2   Simulating Soft Tissue Deformations

Simulation of surgical and interventional medical procedures such as needle insertion requires estimating biomechanical deformations of soft tissue when forces are applied. Because of the complicated geometry of tissue and the wide array of possible forces that can be applied by surgical instruments, closed form solutions for soft tissue displacement fields cannot be computed in general.

Historically, several methods have been developed for discretizing tissue into smaller elements for which the equations of continuum mechanics can be directly applied, and then numerically combining solutions from the discrete chunks into a global solution to obtain a tissue displacement field. The history of offline animation and real-time simulation of deformable objects is summarized in [91]. Here we discuss the mass-spring method and finite element methods, both of which are capable of simulating deformable bodies with complicated geometries and composed of heterogeneous materials.

2.2.1   Mass-Spring Method

Mass-spring methods have been common for simulating a diverse array of human tissues including muscles [207] and blood vessels [45]. In this method, the tissue is defined using a discrete set of virtual point masses, or nodes, that represent the tissue volume. Because each node represents a small volume of tissue around it, this approach is sometimes referred to as the lumped element model (LEM) [49]. Adjacent nodes are connected by massless linear response springs to form a 2-D or 3-D linkage of springs, as shown in figure 2.4. The dynamics of a node j in the mass-spring model is governed by $\mathbf{F}_j = \mu_j \mathbf{a}_j$, where $\mathbf{a}_j$ is the acceleration of node j, μ_j is the mass of node j, and $\mathbf{F}_j$ is the force applied to node j, which includes external forces and internal forces based on spring compression or extension. Viscous forces can also be added. Standard explicit or implicit time integration methods can be used to compute velocity and position of each node j from acceleration for each time step of the simulation [28]. Mass-spring models are relatively easy to implement. However, the arrangement of 1-D springs to define a 2-D or 3-D object has a significant impact on the deformation behavior of the object making it difficult to restrict volume changes or to model isotropic (or pre-defined anisotropic) material properties. Furthermore, there is no direct connection between spring stiffness coefficients and the Young's modulus and Poisson's ratio of continuum mechanics.

Fig. 2.4. A regular mass-spring mesh for a 2-D object. The horizontal and vertical springs resist compression or tension while the diagonal springs are required to resist pure shear strains.

2.2.2 Finite Element Method

The development of the finite element method (FEM) can be traced back to the early 1940's, including key contributions by Alexander Hrennikof and Richard Courant. Finite element methods have been used extensively in the mechanical engineering community to model stiff materials. Unlike the mass-spring method, the finite element method is directly based on the equations of continuum mechanics. The feasibility and potential of using a finite element approach for computer animation was demonstrated by Terzopoulos et al. [200], and Stéphane Cotin et al. made early contributions to finite element modeling for surgery simulation [57].

Details on the finite element method are available in standard texts [227]. The first step of the finite element method is to subdivide the deformable body into a finite set of elements. These elements correspond to a geometric discretization of the object. Field quantities, like displacement, velocity, or acceleration, can be interpolated within each element using *shape functions* specific to the element shape. Finally, the equations of continuum mechanics can be applied to numerically solve for the interactions between the elements.

Geometric Discretization

In FEM, a deformable object is decomposed into a mesh of simple elements, generally triangles or quadrilaterals in 2-D or tetrahedra or hexahedra in 3-D. An extreme point of any element is called a node.

The *reference mesh* G defines the geometry of the deformable object in its reference state, where G is composed using n nodes and m elements. Each node's coordinate is stored in the node coordinate vector $\mathbf{x}$. In 2-D, $\mathbf{x}$ is of dimension $d = 2n$ and each node has 2 displacement degrees of freedom (DOF). A deformation is defined by a displacement vector $\mathbf{u}$, which specifies the displacement of each

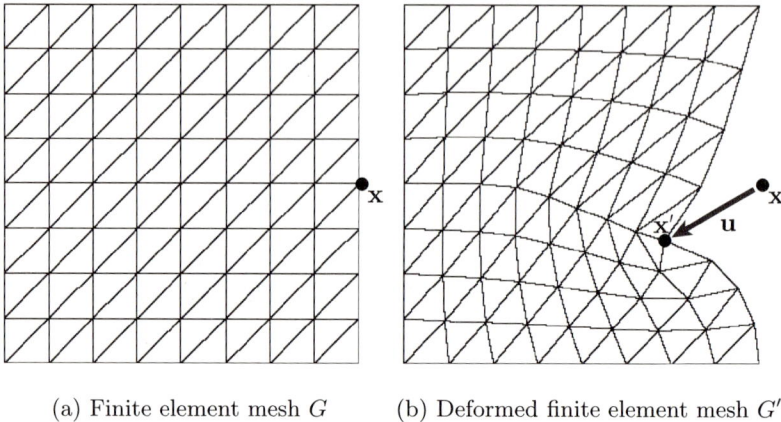

(a) Finite element mesh G (b) Deformed finite element mesh G'

Fig. 2.5. A 2-D finite element reference mesh G composed of triangular elements is shown in (a). The deformed mesh G' is shown in (b). The node at position $\mathbf{x}$ in mesh G is displaced by vector $\mathbf{u}$ to position $\mathbf{x}'$ in deformed mesh G'.

node in mesh G. The *deformed mesh G'* is constructed in the world frame using the displaced node coordinates $\mathbf{x}' = \mathbf{x} + \mathbf{u}$, as shown in figure 2.5.

In continuum mechanics, boundary conditions specify constraints on the deformation of a deformable body. Boundary conditions can be applied at any nodes in the finite element mesh. These include displacement constraints (such as marking a node as fixed, like a node inside a bone) and external forces (due to tool/tissue interactions, like needle insertion). We define external forces in an external forces vector $\mathbf{f}$, with dimension $2n$ in 2-D, where each entry corresponds to a displacement degree of freedom in $\mathbf{x}$.

Interpolation

In continuum mechanics, field quantities such as displacement, velocity, acceleration, and mass density are continuous over the deformable body. Given values for these quantities at n discrete nodes in a finite element mesh, we use *shape functions* (also known as *basis functions*) to interpolate the value of these quantities at any point in the deformable body. Since adjacent elements share common nodes, the field quantities will be continuous over the entire deformable body defined by the mesh. In section 3.3.4, we use linear shape functions within each element, which can be derived using barycentric coordinates [167].

Solving System of Equations

Using a geometric discretization and shape functions, the finite element method provides a framework for calculating the internal stress distribution of elements in the mesh [227]. More specifically, given a vector of external forces acting on nodes of the deformable body, we can compute a vector of node displacements such that the internal forces generated by element stresses balance the external forces.

A simulation computes displacement $\mathbf{u}_i$ as a function of time step i as external forces $\mathbf{f}_i$ change over time. In this discussion, we initially assume the deformable body does not undergo large deformations or rotations and assume a linear relationship between stress and strain (Cauchy strain).

Linear Quasi-static Formulation

The static formulation minimizes the total strain energy over the deformable body to compute its static equilibrium state [227]. Using the finite element method, deformation $\mathbf{u}_i$ at time step i is computed using the formula

$$\mathbf{K}\mathbf{u}_i = \mathbf{f}_i$$

where $\mathbf{u}_i$ is the nodal displacement vector, $\mathbf{f}_i$ is the force displacement vector, and $\mathbf{K}$ is the stiffness matrix based on the material properties of the elements in the mesh defining the deformable object. A quasi-static simulation assumes the deformable object reaches its equilibrium state at each time step.

Real-time visual performance for surgery simulation of the human liver using linear quasi-static FEM was achieved by Stéphane Cotin et al., although the required preprocessing step took 8 hours on a standard PC [58]. This method, for smaller meshes, was also used by DiMaio et al. for modeling force distributions during needle insertion in tissue phantoms [69, 70].

Dynamic Formulation and the Newmark Method

Rather than calculating only static deformations, we can simulate the dynamic behavior of soft tissues by solving for the acceleration, velocity, and displacement of each node for every time step to produce a history-dependent simulation. For a 2-D mesh composed of 3-node triangular elements, the dynamic FEM problem is defined by a system of $d = 2n$ linear equations:

$$\mathbf{M}\mathbf{a}_i + \mathbf{C}\mathbf{v}_i + \mathbf{K}\mathbf{u}_i = \mathbf{f}_i \qquad (2.6)$$

where $\mathbf{M}$ is the mass matrix, $\mathbf{C}$ is the damping matrix, $\mathbf{K}$ is the stiffness matrix, $\mathbf{f}_i$ is the external force vector, $\mathbf{a}_i$ is the nodal acceleration vector, $\mathbf{v}_i$ is the nodal velocity vector, and $\mathbf{u}_i$ is the nodal displacement vector at time step i [227]. The matrices $\mathbf{M}$, $\mathbf{C}$, and $\mathbf{K}$ are defined using the material properties of the elements in the mesh defining the deformable object, which include stiffness, compressibility, Rayleigh damping coefficients, and mass density [227]. Since they are constructed by superimposing the element mass, damping, and stiffness matrices, the number of non-zero entries in each of these matrices is $O(d)$. When a node in the reference mesh is moved or constrained, these matrices must be updated, a process that takes constant time for each DOF.

To solve for $\mathbf{u}_i$ from its time derivatives $\mathbf{v}_i$ and $\mathbf{a}_i$ in the system 2.6, we integrate over time for each time step i. One efficient option is to use the Newmark method [214], which translates the differential system into a linear system of equations with parameters β and γ which are used to solve for displacement $\mathbf{u}_{i+1}$ and velocity $\mathbf{v}_{i+1}$. Let h be the time step duration. Displacement and velocity for the next time step are approximated as:

$$\mathbf{u}_{i+1} = \mathbf{u}_i + h\mathbf{v}_i + (1 - \beta)\frac{h^2}{2}\mathbf{a}_i + \beta\frac{h^2}{2}\mathbf{a}_{i+1}$$

$$\mathbf{v}_{i+1} = \mathbf{v}_i + (1 - \gamma)h\mathbf{a}_i + h\mathbf{a}_{i+1}$$

We consider two solvers: a slower, more accurate solver for planning and a faster solver for interactive simulation. When real-time interactive performance is desired, the value of h is adaptive; it is set using the system clock to the amount of time that has passed since the last iteration was completed.

When more accuracy is required, we set the Newmark method parameters $\beta = 0.5$ and $\gamma = 0.5$ to obtain the implicit system:

$$\left(\mathbf{M} + \frac{h}{2}\mathbf{C} + \frac{h^2}{4}\mathbf{K}\right)\mathbf{a}_{i+1} = \mathbf{f}_{i+1} - \left(\frac{h}{2}\mathbf{C} + \frac{h^2}{4}\mathbf{K}\right)\mathbf{a}_i - (\mathbf{C} + h\mathbf{K})\mathbf{v}_i - \mathbf{K}\mathbf{u}_i$$

$$\mathbf{v}_{i+1} = \mathbf{v}_i + \frac{h}{2}(\mathbf{a}_i + \mathbf{a}_{i+1})$$

$$\mathbf{u}_{i+1} = \mathbf{u}_i + h\mathbf{v}_{i+1} + \frac{h^2}{4}(\mathbf{a}_i + \mathbf{a}_{i+1})$$

Acceleration is obtained by solving the linear system using an iterative numerical method such as Gauss-Seidel or Conjugate Gradient that takes advantage of the sparsity of the matrices. Since $\mathbf{K}$, $\mathbf{M}$, and $\mathbf{C}$ contain only $O(d)$ non-zero entries, the iterative method will take $O(d^2)$ time in the worst case, although typically the number of iterations is much less than d.

For interactive simulation, we avoid solving a linear system by setting the Newmark method parameters to $\beta = 0$ and $\gamma = 0.5$ to obtain an explicit system.

$$\mathbf{u}_{i+1} = \mathbf{u}_i + h\mathbf{v}_i + \frac{h^2}{2}\mathbf{a}_i$$

$$\left(\mathbf{M} + \frac{h}{2}\mathbf{C}\right)\mathbf{a}_{i+1} = \mathbf{f}_{i+1} - \mathbf{K}\mathbf{u}_{i+1} - \mathbf{C}\left(\mathbf{v}_i + \frac{h}{2}\mathbf{a}_i\right)$$

$$\mathbf{v}_{i+1} = \mathbf{v}_i + \frac{h}{2}(\mathbf{a}_i + \mathbf{a}_{i+1})$$

Mass lumping, which approximates the continuous material as a particle system, decouples the system of equations into a set of algebraic equations [173, 225]. For soft materials, mass lumping results in a small loss of accuracy in the dynamics of the object [225]. With mass lumping, each time step requires only $O(d)$ time to compute and does not require any extensive pre-computation.

In most cases, explicit integration is considered inferior to implicit integration because it is unstable for large time steps [28]. However, this instability is most prevalent for stiff materials since the maximum time step length is inversely proportional to the natural frequency of the dynamic system 2.6. Since the natural frequency is small for soft tissues, explicit integration can often be used effectively for these simulations [225].

Nonlinear FEM

The above finite element formulations use Cauchy strain, a linear approximation that loses accuracy for greater deformations. Green's strain, or quadratic strain, correctly handles larger strains and global rotations [155, 167, 173, 226]. Zhuang and Canny and Picinbono et al., in addition to relaxing the quasi-static assumption, also simulate large deformations using quadratic strain, which generates a nonlinear system of equations [173, 226]. To achieve real-time visual performance for reasonably sized meshes, Zhuang uses two key approximations: mass lumping (as described above) and a graded mesh.

To accurately model large deformations, it may also be necessary to take into account the nonlinear elasticity of some materials [25, 216]. Azar et al. develop an offline FEM model of the female breast to track the position of a tumor for a biopsy procedure [25]. Because of the large deformations caused by compression, a piece-wise linear function was used to approximate the nonlinear elasticity of the tissues. Wu et al. use mass lumping and adaptive mesh refinement to achieve real-time performance [216]. A key mathematical limitation of using Green's strain is that it cannot properly handle large compressions; simulated internal forces incorrectly decline when an element is compressed to less than 30% of its material volume [167].

2.2.3   Visualizing 2-D Simulations

The visual feedback of a simulation should mimic the experience of a physician performing the simulated medical procedure [19]. For 2-D simulations of image-guided procedures, this can be performed efficiently using standard computer graphics hardware.

We use a static 2-D image of the tissue as input. As the user of the simulation deforms the tissue, we deform the input image to match the deformations computed in the mesh. We implement this visualization using texture-mapping [99], where the deformed image is constructed by using mesh $\mathcal{G}$ to obtain texture map coordinates for $\mathcal{G}'$. Because 2-D texture-mapping is implemented in hardware on all modern graphics cards, using this method does not substantially penalize the speed of the simulation. In addition to displaying the deformed soft tissue, the simulation also allows the user to selectively overlay clinically relevant information, such as organ outlines, medical instruments, and the target location in the deformable tissue. A sample image of a deforming prostate is shown in figure 2.6.

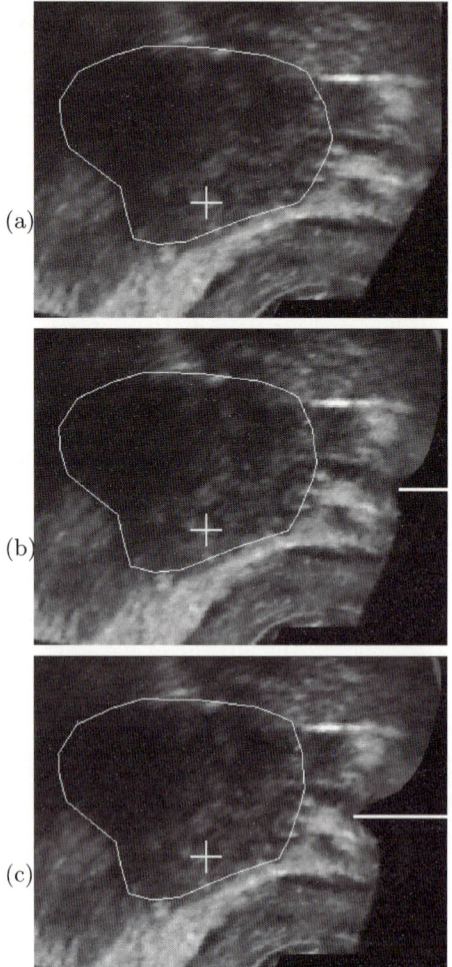

Fig. 2.6. The simulation visualization, which is based on an ultrasound image, is intended to mimic the image that would be seen by a physician during an image-guided procedure. Three frames illustrate the simulated deformations of the prostate (polygonal outline) caused by poking the surrounding tissue from the right. The visualization of each frame is obtained by deforming the single static ultrasound image that was provided as input.

2.3   Conclusion

The methods described in this chapter lay the foundation for the simulations that will be used in chapters 3 and 4 as well as the image registration method in appendix A. This foundation will be used to develop simulations of medical needle insertion and needle steering in deformable soft tissue, which will be used as components of motion planning systems..

As with most modeling problems, there is an inherent trade-off between simulation realism and computational complexity. However, new simulation algorithms are constantly being developed that push the trade-off curve outward and both improve realism and reduce computation time. Ideally, as the capabilities of physically-based simulations improve, these improvement should be directly incorporated into the simulation-based motion planners presented in future chapters.

3 Motion Planning in Deformable Soft Tissue with Applications to Needle Insertion

Minimally invasive medical procedures such as brachytherapy, biopsies, and treatment injections often require inserting a rigid needle to a specific target location inside the body to implant a radioactive seed, extract a tissue sample, or inject a drug. In all cases, the needle tip should be inserted as close as possible to a predetermined target inside soft tissue. Unfortunately, inserting and retracting a needle causes the surrounding soft tissue to displace and deform: ignoring these deformations can result in substantial placement error, as illustrated in figure 3.1. Although real-time imaging such as ultrasound is available during the procedure, it does not provide crisp tissue boundaries and cannot be used to precisely monitor the tissue deformations. Physicians must therefore learn to compensate for the effects of tissue deformations in order to insert the needle to the correct location within the tissue.

We develop a sensorless planning system based on a biomechanical simulation of needle insertion to reduce placement error. Here, "sensorless" refers to a minimalist approach to robotics in which no real-time sensor input is required as the procedure is performed [77]; i.e., no real-time tracking of tissues or the needle is required during the procedure. Our pre-operative planning system combines the simulation described in chapter 2 with an optimization algorithm to compute a needle offset that compensates for tissue deformations to reach a given target location. The planner iteratively tests different insertion locations and depths to compute the optimal needle offset: a sensorless motion plan as illustrated in figure 3.1 right column greatly reduces placement error in simulation.

We apply the system to permanent seed prostate brachytherapy, a minimally invasive medical procedure in which a physician uses needles to permanently implant radioactive seeds inside the prostate that irradiate surrounding tissue over several months. The success of this procedure depends on the accurate placement of radioactive seeds within the prostate gland [62, 178]. For permanent seed brachytherapy, we define seed *placement error* as the Euclidean distance between the desired location specified by the dosimetric plan (the target) and the actual implanted seed location after needle retraction. An experienced physician implanting seeds (without stabilizing needles) in 20 patients achieved average

R. Alterovitz and K. Goldberg: Motion Planning in Medicine, STAR 50, pp. 27–44, 2008.
springerlink.com © Springer-Verlag Berlin Heidelberg 2008

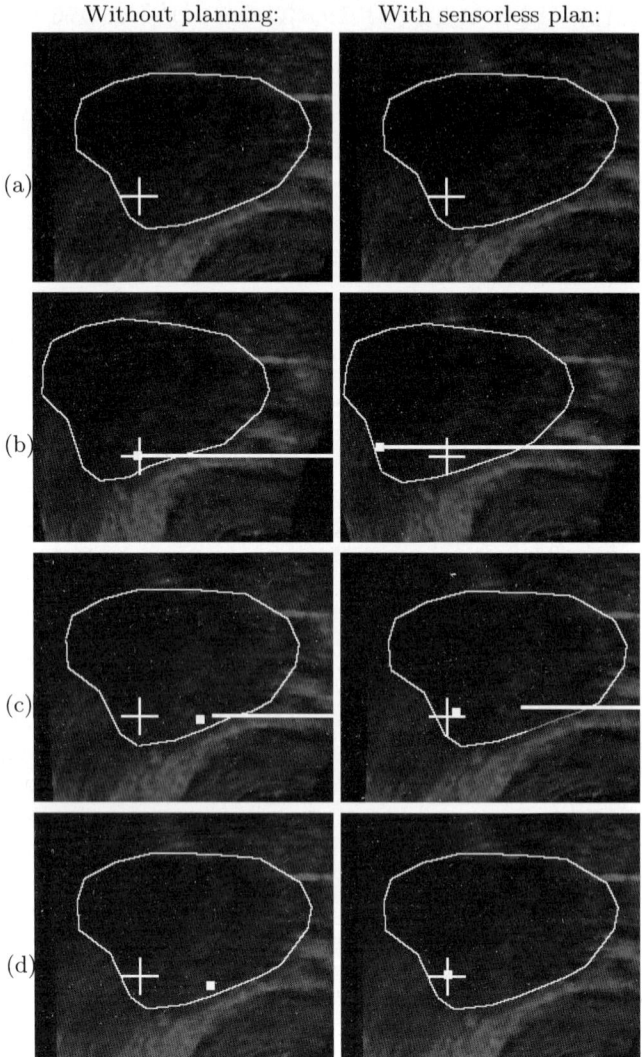

Fig. 3.1. Four vertical frames illustrate brachytherapy needle insertion based on deforming an ultrasound image of the human prostate using simulation. The left column shows results without planning, producing substantial placement error. The right column shows results with the sensorless plan, with minimal placement error. The target implant location is indicated in all frames with a cross fixed in the world frame. Frame (a) outlines the undeformed prostate. In Frame (b), the needle is inserted and the radioactive seed (small square) is released at the needle tip. In Frame (c), the needle is retracted. Frame (d) indicates the resulting placement error, the distance between the target and resulting actual seed location. Without planning, placement error is substantial: 26% of the prostate diameter, resulting in damage to healthy tissue and failure to kill cancerous cells. With sensorless planning, shown in the bottom image of Frame (d), placement error is negligible in this simulation.

placement errors of 0.47 cm in depth and 0.22 cm in height for an average placement error of 0.63 cm, a substantial error of 21% of average prostate diameter (3 cm) [196]. Real-time ultrasound imaging is used during the procedure to help guide each needle along a straight path and to verify the depth of the needle tip in the world frame. However, the imaging cannot effectively be used to compensate for deformations because it does not include crisp markers with known positions inside the soft tissues, as shown in figure 1.4. Tissue deformations during needle insertion and retraction contribute to placement error during brachytherapy [178, 196], as illustrated in figure 3.1. In this chapter, we describe our sensorless planning approach to reduce placement error without relying on real-time imaging.

In section 3.3, we introduce a simulation of needle insertion in deformable tissue that can be used interactively for physician training or offline for procedure planning. In section 3.4, we use the simulation as a component of the planning system that computes needle insertion offsets to compensate for the effect of tissue deformations. In section 3.5, we apply the simulator and planner to minimize the placement error of radioactive seed implants for prostate brachytherapy.

To define the anatomy and clinical target, the simulation requires a pre-procedure medical image. Many imaging methodologies, such as ultrasound, display a 2-D planar cross-section of the human body. MR images, which may contain multiple planar slices composing a 3-D volume, have an inter-slice distance significantly greater than the diameter of a medical needle. Hence, in this chapter we restrict needle motion to a 2-D cross-section of the patient anatomy.

3.1   Sensorless Planning and Needle Insertion

In robotics, sensorless planning algorithms, pioneered by Mason and Erdmann in the 1980's [77], have been developed to position and orient mechanical parts using parallel jaws [43, 94], vibrating surfaces [37], single joint robots over conveyor belts [5], and squeeze and roll primitives for micro-scale parts [156]. For needle placement planning using rigid needles, our goal is to model and compensate for mechanical response before actions are performed.

Medical needle insertion procedures may benefit from the more precise control of needle position and velocity made possible through robotic surgical assistants. Surveys of recent advances in medical robotics have been written by Taylor and Stoianovici [199] and Cleary et al. [56]. Dedicated hardware for needle insertion is being developed for stereotactic neurosurgery [151], MR compatible surgical assistance [50, 68], and prostate biopsy and therapeutic interventions [83, 84, 172, 186].

When real-time sensor data such as MR or X-ray imaging is available during needle insertion procedures and the target and relevant obstacles are all discernible in the images, robotic control algorithms can be used to steer the needle to the desired target. Shi et al. developed an image-guided robotic system containing a needle as an end-effector that uses real-time X-ray imaging to track a target and send its position to a control system [192]. The needle's tip position is computed using forward kinematics and the control system repeatedly updates the insertion path of the needle tip to a straight line path to the target.

When real-time sensor data is unavailable or unreliable, sensorless planning based on pre-operatively predicting the effects of tissue deformations can be applied. Azar et al. use a piece-wise nonlinear finite element model to track the position of a tumor during breast compression before a breast cancer biopsy [25]. Recent work has addressed planning local trajectories in deformable tissue for flexible needles with symmetric tips by translating and rotating the base [72, 92] and steerable bevel-tip needles that can be controlled by rotating the bevel [10, 208]. In this chapter, we explicitly use simulation of insertion of rigid needles into deformable tissues to plan needle procedures without real-time sensor input [18].

3.2   Problem Formulation

As illustrated in figure 3.2, we consider a 2-D slice of tissue in the yz plane. At time $t = 0$, the tissue is at rest (undeformed). The target is denoted by a point $\mathbf{g} = (y_g, z_g)$ in the world frame at time $t = 0$.

A needle motion plan is defined by a control vector $\mathbf{X}$. We define $\mathbf{X} = (y_r, z_r)$ where y_r is the "insertion height" and z_r is the "insertion depth." A needle insertion procedure consists of inserting the needle at height y_r to a depth z_r, implanting a radioactive seed at this release point, retracting the needle, and waiting for steady-state. For simplicity, we assume that the needle moves parallel to the z-axis and that the coordinate system of the needle and the coordinate system of the tissue are identical. The location of the seed in the world frame after retraction is denoted by $\mathbf{p} = (y_p, z_p)$.

Due to the effects of tissue deformation, $\mathbf{X} \neq \mathbf{p}$. We measure seed placement error using the Euclidean distance between the final seed location $\mathbf{p}$ and the target location $\mathbf{g}$:

$$\varepsilon = \|\mathbf{p} - \mathbf{g}\| .$$

The physically-based simulation is essentially a function whose input is a motion plan $\mathbf{X}$ and whose output is the final seed placement $\mathbf{p}$. Hence, we specify the simulation as a function $S(\mathbf{X})$ that returns the final seed placement:

$$\mathbf{p} = S(\mathbf{X}).$$

For a given target point $\mathbf{g}$ inside soft tissue, the motion planning problem is to compute a plan $\mathbf{X}$ that minimizes placement error. To compute an optimal motion plan $\mathbf{X}^*$, we use the simulation as a function in the optimization:

$$\mathbf{X}^* = \arg\min_{\mathbf{X}}(\varepsilon) = \arg\min_{\mathbf{X}}(\|S(\mathbf{X}) - \mathbf{g}\|). \tag{3.1}$$

During planning, we restrict the range of the parameters of the motion plan $\mathbf{X}$ to clinically feasible values based on anatomical constraints. We restrict the y_r to the region of skin where the needle can be feasibly and safely inserted, $y_r \in (y_{min}, y_{max})$. We define z_{max} as the maximum medically feasible needle insertion depth.

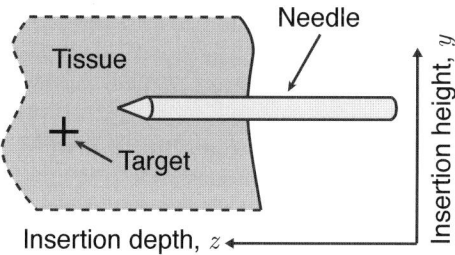

Fig. 3.2. Slice of deformable tissue in the yz plane. The needle is inserted from right to left parallel to the z-axis, causing the tissue to deform.

3.3   Simulating Needle Insertion

In this section, we introduce a simulation of the insertion and retraction of a thin, rigid, symmetric tip needle in a 2-D slice of deformable soft tissue. We use a finite element method (FEM) to compute the deformations of soft tissues when forces are applied by the needle. Rather than calculating only static deformations, we simulate the dynamic behavior of soft tissues by solving for the acceleration, velocity, and displacement of each node for every time step to produce a history-dependent simulation.

As in related work [58, 70], we approximate soft tissues as linearly elastic, isotropic materials (Cauchy strain). Tissue may be inhomogeneous but must be fully connected with no gaps between different tissue types. We do not model slip between tissue types or physiological changes that result from needle insertion, such as edema (tissue swelling). As computation speed improves and biomechanics experiments provide more nonlinear tissue properties, this simulation can serve as foundation that can be extended to incorporate more complex tissue models.

The simulation uses a model of needle insertion based on a set of scalar parameters including needle friction, sharpness, velocity, and insertion location. These parameters can be selected, within limits, by the physician to improve placement accuracy. This model allows us to produce an interactive simulation and analyze the sensitivity of current medical methods to these parameters [16, 17, 19].

3.3.1   Background on Needle Insertion Modeling and Simulation

Abolhassani et al. provide a survey of models and simulations of needle insertion [3]. Simulating needle insertion requires a model of the forces exerted by the needle on soft tissue. Okamura, Simone, and O'Leary measured needle insertion forces during robot-assisted percutaneous therapy and separated the forces into distinct components: tissue stiffness forces, a cutting force at the needle tip, and frictional forces along the needle shaft [168, 194]. Kataoka et al. separately measured cutting and frictional forces during needle insertion into a canine prostate [117]. We include these force components in our model of needle insertion.

DiMaio and Salcudean extracted needle insertion force profiles from camera images of needle insertion in an artificial tissue phantom [69, 70]. They used a quasi-static finite element method to replicate tissue phantom experiments in simulation, and extract a force distribution, which they modeled with a parameterized surface. Directly integrating these force profiles into a quasi-static finite element simulation of needle insertion results in a simulation with extremely fast update rates (500Hz), which is sufficient for both visual and haptic feedback. However, the method for extracting force profiles cannot be directly performed *in-vivo* for living tissues.

Alterovitz et al. used simulation to show that needle insertion velocity has an effect on placement error [17], and Heverly et al. developed detailed models and physical experiments measuring this effect [100]. Recent work has also provided physical measurements and models of the bending of needles during insertion [2], which may significantly affect procedure outcome for thinner needles.

Setting accurate parameters for tissue material properties is also important for realistic simulation of needle insertion. Krouskop et al. estimated the elastic modulus for prostate and breast tissue using ultrasonic elastography [126]. Recent work has estimated nonlinear tissue property parameters [46, 115].

To simulate needle insertion, needle cutting and frictional forces are applied at nodes of the finite element mesh. DiMaio and Salcudean relied on node snapping, which moves the closest mesh node to the needle path in the world frame [69, 70]. Nienhuys et al. proposed mesh refinement to mitigate the discretization error caused by node snapping [164]. These methods incur an error in the location of applied needle forces that is dependent on the tissue mesh density. The method we present here uses mesh modification to move nodes in the reference mesh to the needle tip and along the needle shaft. The benefits of this approach include that no extra elements need to be created and the path cut by the needle is directly encoded within the reference mesh.

3.3.2 Input Anatomy Model

We represent the anatomy geometry (i.e. the tissues relevant to the simulation) using a finite element mesh. The input required for our geometric model includes a bitmap image of a 2-D slice of tissue and a segmentation of the tissue types in the image using polygons. Based on the polygonal segmentation, we automatically generate a finite element *reference mesh* $\mathcal{G}$ composed of n nodes and m triangular elements in a regular right triangle mesh or using the constrained Delaunay triangulation program *Triangle* [191]. Each node's coordinate is stored in the node coordinate vector $\mathbf{x}$. In 2-D, each node has 2 degrees of freedom (DOF) so $\mathbf{x}$ is of dimension $d = 2n$.

To compute tissue deformations, the model must also include tissue material properties, boundary conditions for the finite element mesh, and needle properties. For each segmented tissue type, the model requires as input the tissue material properties (i.e. the Young's modulus and Poisson ratio for linearly elastic materials). Each element in the mesh may be assigned unique material properties, which allows for the simulation of multiple tissue types in one mesh. Mesh nodes defining elements inside bones are constrained to be fixed. A boundary condition of either free or fixed must be specified for each node on the perimeter of the finite element mesh. Needle properties that must be specified include the cutting force (force required to cut a unit length of tissue) and the static and kinetic coefficients of friction between the tissue and needle. We discuss a particular anatomy model, for the prostate, in section 3.5.

3.3.3   Simulation Output

The simulation computes mesh deformations that estimate the tissue's response to the needle over time. The deformation is defined by a displacement vector $\mathbf{u}$, which specifies the displacement of each node in mesh $\mathcal{G}$. The *deformed mesh* $\mathcal{G}'$ is constructed in the world frame using the displaced node coordinates $\mathbf{x} + \mathbf{u}$. The simulation computes the displacement $\mathbf{u}_i$ as a function of time step i. Using a fixed time step duration h, we obtain simulated deformations for times $t=hi$, $i \geq 0$.

3.3.4   Simulating Needle Procedures

Without loss of generality, we set the coordinate axes of the world frame so that the needle is inserted along the z axis. In 2-D, the y-axis corresponds to needle insertion height. Once the needle is in contact with tissue, we assume the needle's y-coordinate is fixed and it only moves parallel to the z-axis. Needle insertion corresponds to increasing depth z, as shown in figure 3.2.

Rather than modeling the needle as a distinct meshed object, we instead model the needle implicitly by applying needle insertion forces to the surrounding soft tissue. We then use a finite element method (FEM) with Newmark time integration, as described in section 2.2, to compute the displacement vector $\mathbf{u}$ for soft tissues due to the forces applied by the needle.

The needle exerts force on the tissue at the needle tip, where the needle is displacing and cutting the tissue, and along the shaft due to friction [194]. We model these forces and apply them as the force vector $\mathbf{f}_i$, which we update at every time step of the simulation. This implicit method for representing the needle facilitates real-time interactive performance since no expensive collision detection between the needle and soft tissue is required.

We apply the needle insertion forces as boundary conditions on elements in the mesh. Since the needle may be inserted at any location, it is necessary to modify the reference mesh in real-time to ensure that element boundaries are present where the tip and friction forces must be applied. To apply the tip force, a node is maintained at the needle tip location during insertion. To apply the friction forces, a list of nodes along the needle shaft is maintained and these shaft nodes are constrained to only move along the z-axis.

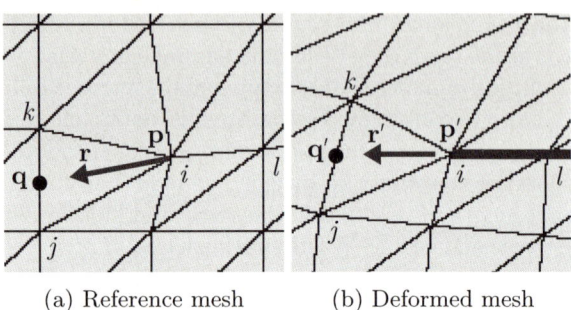

(a) Reference mesh   (b) Deformed mesh

Fig. 3.3. The needle is in the interior of the mesh with needle tip node $c = i$ at point $\mathbf{p}$

Cutting at the needle tip

By default, inserting the needle causes the tip to push tissue but not cut it. This corresponds to a displacement of the needle tip node c in the world frame, but no change in the reference mesh. At every time step, we measure the force f_c applied by the tissue onto the needle tip. We define f_b as the magnitude of the force required to cut a length b of tissue where f_b depends on the sharpness of the needle. For each time step in which $f_c \geq f_b$, we modify the reference mesh to move the needle tip node a distance b.

We illustrate this mesh modification in figure 3.3. Let point $\mathbf{p}$ be the location of the needle tip node c in the reference mesh. The needle tip at node $c = i$ is moving horizontally to the left in the world frame as shown by the vector $\mathbf{r}'$ in figure 3.3(b). This vector is linearly transformed [227] to the reference mesh in figure 3.3(a) and is denoted by $\mathbf{r}$. We move the needle tip node a distance b along $\mathbf{r}$ in the reference mesh to a new point $\mathbf{p} + b\mathbf{r}$. After each time step in which the needle cuts tissue, $\mathbf{p}$ moves closer to $\mathbf{q}$ where $\mathbf{q}$ is the projection of the vector $\mathbf{p} + b\mathbf{r}$ onto the opposite segment (j, k). To maintain a planar mesh with non-overlapping elements, we periodically select a new tip node. When the Euclidean distance from node l (the first node on the needle shaft behind the tip node) to node i is more than twice the distance from node i to point $\mathbf{q}$, node i is added to the needle shaft: the z-component of node i is freed and returned to its original value and the node is constrained to lie on the needle axis by fixing its y-component degree of freedom. The closer of node j or k is moved to $\mathbf{p} + b\mathbf{r}$ and is defined as the new tip node c. Key frames from a simulation using this type of mesh modification are shown in figure 3.4.

To maintain simulation stability, it is necessary to maintain a topologically valid planar mesh in which all elements have strictly positive area. Using the mesh modification above on a sparse mesh, the tip node may move such that triangle (i, l, h) has negative area, as shown in figure 3.5. For this to occur, the y-component of $\mathbf{r}$ must change sign twice over the span of just two element edges. We can avoid this situation by using a finer mesh that prevents the necessary conditions for negative area triangles from occurring, or by using the method

Reference mesh: Deformed mesh:

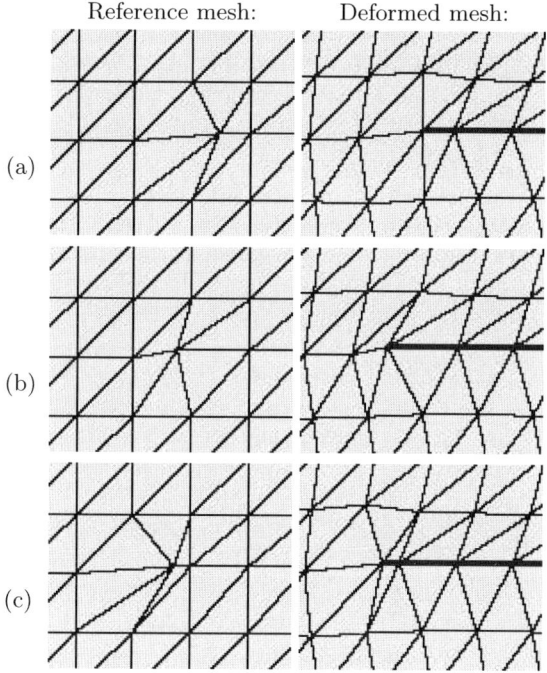

(a)

(b)

(c)

Fig. 3.4. The needle tip is inserted to the left in (a) through (c). The tip node is moved onto the shaft in (c) and the next tip node is selected.

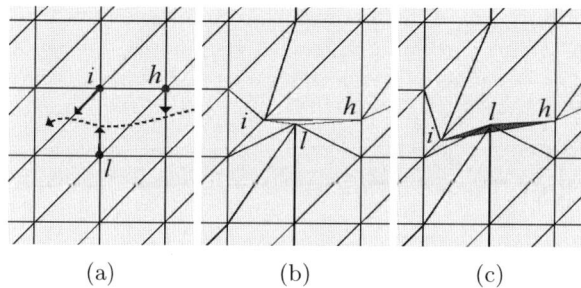

(a)   (b)   (c)

Fig. 3.5. A portion of a reference mesh with a needle path (the dotted line) is shown in (a) with tip node i and shaft nodes l and h. As the tip node i moves downward in (b) and (c), triangle (i, l, h) becomes degenerate. This situation can be corrected using local re-meshing [166] or avoided by using a finer mesh.

proposed by Nienhuys and van der Stappen [166] to efficiently modify the mesh using local edge flips to maintain a valid Delaunay triangulation. We used a sufficiently fine mesh such that local re-meshing was never required and the simulation was sufficiently fast for interactive performance.

Friction along the needle shaft

Our stick-slip approach to modeling static and kinetic friction between the needle shaft and the tissue is based on the friction model of Baraff and Witkin [28]. When the tangential velocity of a node along the needle shaft and the velocity of the needle are equal to within a small epsilon threshold, then static friction is applied: the node is attached to the needle and moves at the same velocity along the z-axis. When the tangential force required to attach the node to the needle exceeds a slip force threshold, then the node is freed to slide along the needle shaft and a dissipative force is applied.

Seed implantation

At any time during needle insertion, a seed can be implanted at the location of the needle tip $\mathbf{s} = \mathbf{p}$. We assume that the seed does not cut tissue, so, after it is implanted, the seed moves in the world frame with the deforming tissue that surrounds it but its coordinate in the reference mesh remains fixed. To maintain the seed at a fixed position in the reference mesh as the mesh is modified, we store in memory the mesh element e containing $\mathbf{s}$. When any node j of element e is moved in the reference mesh during the simulation, we update e by examining each triangle containing node j and checking if point $\mathbf{s}$ is in that triangle using the zero-winding rule [99]. By storing the surrounding element of $\mathbf{s}$ in the reference mesh, we can efficiently compute the location $\mathbf{s}'$ of the seed in the world frame using the shape functions of e for the deformed mesh [227].

Needle retraction

During needle retraction, a tip node is not maintained since no cutting force is required. When the needle retracts past a node on the shaft, that node is removed from the shaft node list. Friction is applied on all the shaft nodes as during insertion.

3.3.5 Simulation Visualization

The visual feedback of the simulation is intended to mimic the experience of a physician performing a needle insertion procedure with ultrasound image guidance [19]. We use a static 2-D image of the prostate as input. As the user of the simulation inserts the needle, we deform the input image to match the deformations computed in the mesh. We implemented this visualization using texture-mapping [99], where the deformed image is constructed by using mesh $\mathcal{G}$ to obtain texture map coordinates for $\mathcal{G}'$. Because 2-D texture-mapping is implemented in hardware on all modern graphics cards, using this method does not substantially penalize the speed of the simulation. In addition to displaying the deformed soft tissue, the simulation also allows the user to selectively overlay clinically relevant information, such as organ outlines, the needle, and the target location in the deformable tissue. A screen capture from the simulation for the prostate brachytherapy application is shown in figure 3.6.

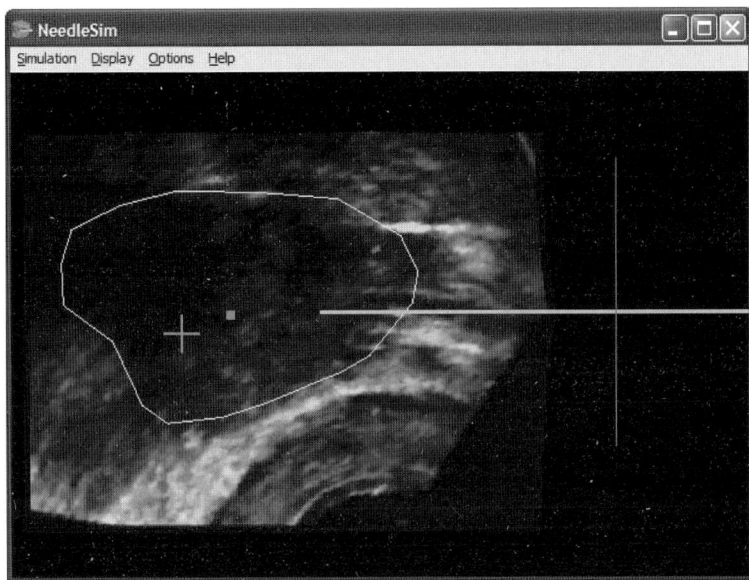

Fig. 3.6. The simulation user interface, which is based on an ultrasound image, is intended to mimic the experience of a physician performing brachytherapy. The physician interactively guides the needle using a mouse and implants seeds (small squares). Tissue deformations and seed locations are predicted and displayed. The implantation error is the distance between the seed and its target (cross) after needle retraction.

3.4   Motion Planning for Needle Insertion

3.4.1   Method Overview

Given a target point $\mathbf{g}$, the goal of needle insertion planning is to find an optimal motion plan $\mathbf{X}^*$ that minimizes placement error $\varepsilon = \|\mathbf{p} - \mathbf{g}\|$, where the final seed implant location $\mathbf{p}$ is a function of the plan $\mathbf{X}$. Because the relationship between $\mathbf{p}$ and $\mathbf{X}$ cannot be defined as a closed-form equation, the optimal $\mathbf{X}^*$ cannot be computed analytically. Our algorithm efficiently uses the simulation as a function in an optimization algorithm to compute the optimal motion plan $\mathbf{X}^*$.

3.4.2   Planning Problem Formulation

The planning algorithm's inputs and outputs are defined by:

Input:

- Needle insertion simulator S (as defined in section 3.3)
- $\mathbf{g}$: Target coordinate in the tissue
- (y_{min}, y_{max}): Range of feasible insertion heights
- z_{max}: Maximum feasible insertion depth

- v: Needle speed during insertion and retraction
- h: Simulation time step

Output:

- $\mathbf{X}^*$: Optimal motion plan that minimizes placement error

A naïve planner that ignores tissue deformations would set $\mathbf{X} = \mathbf{g}$. If tissue deformations occur, the naïve plan will not reach the specified target, as shown in simulation in figure 3.1 left column.

To estimate the optimal plan $\mathbf{X}^*$, the planner computes an offset from $\mathbf{g}$ for both the insertion depth and height. The offset for needle insertion depth is necessary because tissue in front of the needle tip is compressed during insertion; the needle must be inserted deeper than z_g to compensate for this compression. The offset for insertion height is necessary since organs or glands (such as the prostate) may rotate during needle insertion. For example, if the needle is inserted near the bottom of the prostate, the gland will rotate clockwise because it is composed of a stiffer tissue than the surrounding soft tissue, as shown in figure 3.7. Hence, the needle must be inserted higher to compensate for its deflected path through the prostate.

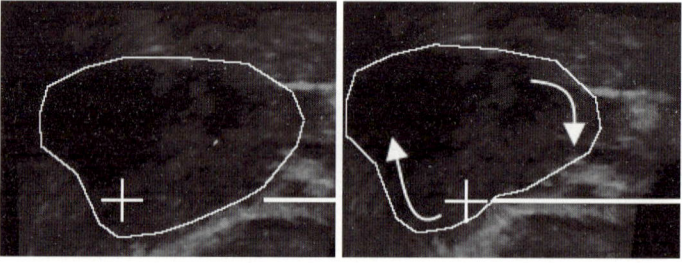

(a) Needle approaches prostate   (b) Prostate rotated by needle

Fig. 3.7. When the needle pushes against the lower half of the prostate from the right, the prostate rotates clockwise because it is stiffer than the surrounding tissue. This rotation can lead to significant changes in the optimal needle insertion height.

3.4.3   Planning Algorithm

We formulate the motion planning problem as an optimization problem, as given in equation 3.1, where $\mathbf{X}$ has 2 degrees of freedom, y_r and z_r. To computationally accelerate the optimization, we consider two one-dimensional problems. First, we implement an algorithm that, given an insertion height y_r, computes the optimal insertion depth z_r:

$$z_r^*(y_r) = \arg\min_{z_r}(\|(S((y_r, z_r)) - \mathbf{g}\|). \tag{3.2}$$

Then, we implement an algorithm that optimizes y_r and uses the first algorithm to implicitly compute z_r^* for each candidate y_r:

$$y_r^* = \arg\min_{y_r}(\|(S((y_r, z_r^*(y_r))) - \mathbf{g}\|). \tag{3.3}$$

Equation 3.2 can be solved efficiently by noting that it is not necessary to fully simulate needle retraction for each candidate plan $\mathbf{X}$. Let k be the node at the needle tip at the time of seed implantation. Since we model tissues as elastic, the displacement $\mathbf{u}_{\mathbf{k}j}$ from system 2.6 will be 0 for all iterations j after the needle has been retracted and steady state is reached. Hence, the location in the world frame of the release point $\mathbf{X} = (y_r, z_r)$ after needle retraction will be $\mathbf{x_k} + \mathbf{u}_{\mathbf{k}j} = \mathbf{x_k}$. Since we assumed that seeds do not cut tissue, the final seed location is $\mathbf{p} = \mathbf{x_k}$ and the placement error is $\varepsilon = \|\mathbf{x_k} - \mathbf{g}\|$, where $\mathbf{x_k}$ is the reference mesh coordinate of the node k at the needle tip when it reaches the release point $\mathbf{X} = (y_r, z_r)$ in simulation. An implication of this is that we can compute the optimal z_r^* in equation 3.2 by running a single simulation of needle insertion from $z_r \leq 0$ until $z_r = z_{max}$. At each time step we compute ϵ in $O(1)$ computation time and record z_r^* for the lowest ϵ. This method is guaranteed to find the optimal z_r^* (within the resolution of the time steps) regardless of the convexity properties of equation 3.2.

Solving equation 3.2 using this approach requires computing $z_{max}/(vh)$ simulation time steps, each requiring $O(d)$ time (or slower if a more accurate FEM model or solver is used) as described in section 3.3. Since the needle tip will move a distance vh each time step, the resolution of z_r^* is vh. A small time step h is desirable to improve the resolution of z_r^*, but the number of time steps required to compute the optimal insertion depth z_r^* grows as h decreases.

Solving equation 3.3 is difficult because derivative values are not available and the function is not guaranteed to be convex. In general, an approximate minimum can be found using a grid search over $y_r \in (y_{min}, y_{max})$. However, equation 3.3 will be unimodal (strictly quasiconvex) near the minimum in cases for which it is not possible to insert the needle at different heights and still reach the same point in the reference mesh of the tissue. Although this property is not guaranteed, it holds for most feasible targets in our simulation that are not adjacent to a tissue type boundary. In such cases, we use a line search method, golden section search [31], to find the optimal y_r^*. Golden section search, a variant of the Fibonacci search that requires fewer function evaluations, does not require derivative information (which is not available in the simulation) and convergence is guaranteed.

3.5   Application to Brachytherapy Cancer Treatment

We apply the simulation and planning framework to brachytherapy for treating prostate cancer. Figure 3.1 provides a simulated case study showing that deformations can produce significant errors in final seed placements during prostate brachytherapy. Placement error should be minimized to achieve the desired radioactive dose distribution.

During permanent seed prostate brachytherapy, roughly 20 stiff needles are each loaded with multiple seeds separated by spacers. As the needle is retracted, the "train" of seeds and spacers are released in the prostate. In this chapter, we only address placement of the first seed in the train and ignore the remaining seeds in each needle. Each needle is fully retracted before the next is inserted. Hence, we assume each needle insertion and seed implantation procedure is independent. Unlike needles, we assume seeds do not cut tissue. Hence, a seed will move only when the surrounding tissue deforms, which satisfies our assumption that an implant moves with the surrounding deforming tissue. Also, a metal block containing approximately 50 holes at fixed coordinates is used by the physician to guide each needle during brachytherapy needle insertion. We relax the discrete insertion coordinate restriction and allow the insertion height y_r to vary continuously, which allows for better minimization of placement error but will require new hardware in medical practice.

3.5.1 Simulation Implementation

Our anatomy model of the prostate is based on data obtained in the operating room at the UCSF Comprehensive Cancer Center from a patient undergoing brachytherapy treatment for prostate cancer. An ultrasound video was recorded using an ultrasound probe in the sagittal plane, as shown in figure 1.4. The first frame of the ultrasound video was segmented by a physician from the UCSF Comprehensive Cancer Center. The segmentation was used to manually generate a mesh composed of $n = 676$ nodes and $m = 1250$ triangular elements for a 3.5cm diameter prostate and surrounding fatty tissue. The ultrasound image also served as the texture map image for the simulator. The boundary of the mesh is defined by a square for which the right face (where the needle is inserted) is free, the bottom face corresponding the trans-rectal ultrasound probe is rigid, and the other two faces are also marked rigid.

The Young's modulus E and Poisson ratio ν are set based on the results of Krouskop et al. to $E = 60$kPa and $\nu = 0.49$ for the prostate and $E = 30$kPa and $\nu = 0.49$ for the surrounding fatty tissue [126]. Needle properties are treated as variables that can be set in the user interface of the simulation. To set default values, we compared the output of the simulation with the ultrasound video and set unknown simulation parameters so that the simulation output closely matched the ultrasound video. UCSF clinicians comparing the two image sequences judged them as highly similar.

The simulator was implemented in C++ using OpenGL for visualization and tested on a 750MHz Pentium III PC with 256MB RAM. For a model with 1250 triangular elements, the simulator responds at the rate of 24 frames per second, sufficient for visual feedback (but not fast enough for haptic control). When executed in planning mode, we assume the needle is inserted at a constant velocity of 0.5cm/sec and use a fixed simulation time step of $h=1/30$ seconds.

The visual feedback of the simulation is intended to mimic the experience of a physician performing brachytherapy. When executed in interactive simulation mode, a physician can guide the needle and implant seeds using a mouse, as shown in figure 3.6. We believe this output can be useful for physician training [19]. The interactive simulation runs on standard PC's running Windows 2000 or XP.

3.5.2 Sensorless Planner Results

To test planner performance, we selected 12 sample points inside the prostate, shown by the crosses in the figure 3.8. We applied golden section search in the range $y_r \in (y_g - 0.2\text{cm}, y_g + 0.2\text{cm})$ with tolerance 0.01cm for each target. Without planning, the average error was 0.59cm (17% of prostate diameter) with a standard deviation of 0.10cm. Using our planner, the average error was reduced in simulation to 0.002cm (0.06% of prostate diameter) with a standard deviation of 0.004cm. The average time to compute the optimal plan $\mathbf{X}^*$ per target was 98 seconds.

We examine in detail the planner results for the example in figure 3.1. The target is located at $\mathbf{g} = (1.50\text{cm}, 3.00\text{cm})$. Using the "default" plan $\mathbf{X} = \mathbf{g}$ with the zero offsets, the seed is implanted at $\mathbf{p} = S(\mathbf{g}) = (1.41\text{cm}, 2.21\text{cm})$, a placement error of $\varepsilon = 0.79\text{cm}$ (23% of the prostate diameter).

In figure 3.9, we plot the placement error $\varepsilon(\mathbf{X} = (y + g, z_r))$ for insertion at the target height $y_r = y_g = 1.5\text{cm}$. The placement error for the "default" plan can be seen at $z_r = z_g = 3.0\text{cm}$. The error in the depth coordinate is caused primarily because the tissue in front of the needle tip is being compressed before it is cut. Inserting the needle deeper than the target depth decreases the error. If insertion height is held constant at $y_r = y_g$, placement error can be reduced by 82% to only $\varepsilon = 0.14\text{cm}$ (4% of prostate diameter) by inserting to a depth of $z_r^* = 3.84\text{cm}$.

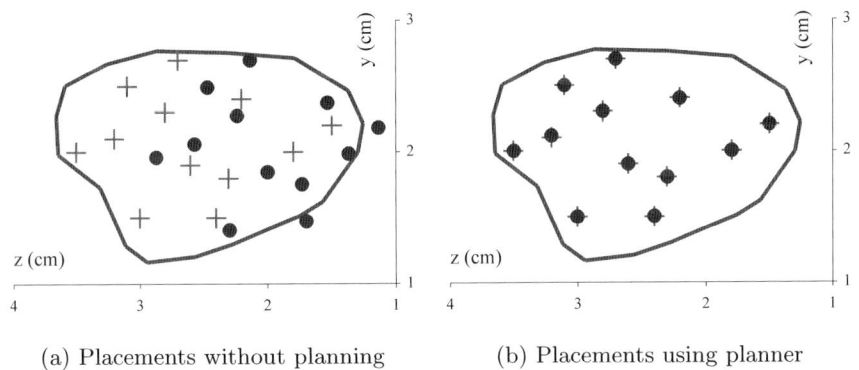

(a) Placements without planning   (b) Placements using planner

Fig. 3.8. Twelve sample points were selected as targets marked "+" inside the prostate. Actual seed placements using simulation are marked "•". Lack of planning results in major placement errors averaging 20% of the prostate diameter (a), which will lead to a poor radioactive dose distribution. Seed placement error was negligible using the planner (b).

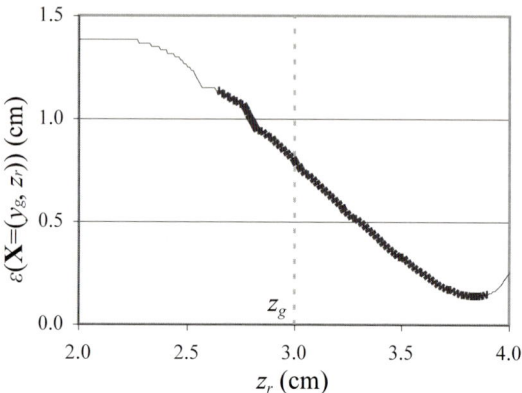

Fig. 3.9. Needles should generally be inserted deeper than the target depth to compensate for tissue deformations and minimize placement error. The bold portion of the line denotes feasible seed placements inside the prostate.

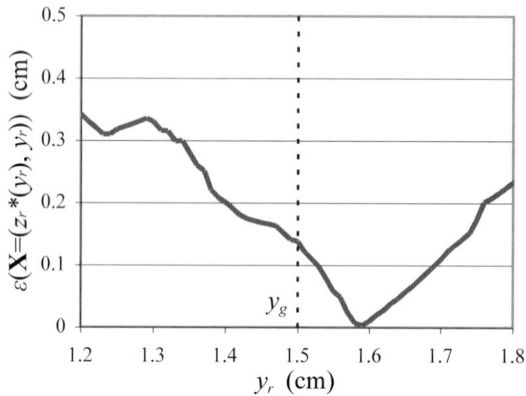

Fig. 3.10. The motion planner computes the optimal insertion height y_r and corresponding optimal depth z_r^* to minimize placement error ϵ. Placement error is minimized for $(y_r, z_r) = (1.59\text{cm}, 3.80\text{cm})$.

In figure 3.10, we plot the optimal surface $\varepsilon(\mathbf{X} = (y_r, z_r^*(y_r))$. The golden section search described in section 3.4.3 efficiently finds the minimum of this surface to determine $\mathbf{X}^*$ with $\varepsilon^* = 0.003\text{cm}$ (0.09% of prostate diameter) by inserting at height $y_r^* = 1.59\text{cm}$ to a depth $z_r^* = 3.80\text{cm}$.

3.6   Conclusion and Open Problems

To facilitate physician training and pre-operative planning for medical needle insertion procedures, we introduced in this chapter a needle insertion motion planning system based on an interactive simulation of the insertion of rigid needles into soft tissue.

The first component of this system is a physically-based, dynamic simulation of needle insertion in soft tissues. In this chapter, we introduce a simulation that uses a 2-D finite element model of soft tissue and a reduced set of scalar parameters such as needle friction, sharpness, and velocity. The simulation modifies the mesh to maintain element boundaries along the needle shaft and models the effects of cutting at the needle tip and frictional forces along the needle shaft. The simulation achieves 24 frames per second for 1250 triangular elements on a 750Mhz PC. The speed of the simulation is determined primarily by the scalability of the finite element method system solver. The computation time per time step will not rise during the simulation because no new nodes are added, and it may in fact decrease as the y-axis DOF of some nodes are lost when they are constrained along the needle shaft.

Using texture-mapping, the simulation provides visualization comparable to ultrasound images that the physician would see during the procedure. A screen capture of the software is shown in figure 3.6. To facilitate physician training, the simulation can be run interactively using different patient anatomies and tissue properties. We hope that needle insertion simulation can be used for physician training as an alternative to the limited and expensive mechanical models currently in use.

We applied our simulation from Chapter 2 as a component of a sensorless planning system for needle insertion procedures. Our sensorless planning method computes needle offsets to minimize needle placement error by compensating for predicted tissue deformations. The approach combines numerical optimization with soft tissue simulation. The effectiveness of the planner *in-vivo* will be dependent on the accuracy of the simulation of tissue deformations that occur during needle insertion for a specific patient.

Recent work has begun exploring 3-D simulation of needle insertion which enables more accurate representation of anatomy and the simulation of needle insertion outside a single imaging plane. Nienhuys et al. used 3-D mesh refinement [165] and Goksel et al. used mesh modification [93] in order to apply needle insertion forces at mesh nodes. However, current 3-D simulations do not perform at interactive or real-time rates, do not offer guarantees on mesh stability throughout the simulation, and do not provide visualization capabilities useful for physician training. The simulation method we presented here considers tissue deformations in 2-D, providing graphical visualization via texture-mapping and offering guarantees on simulation speed and stability that are necessary for physician training and efficient automated planning. Incorporating these features into 3-D simulation is an area of active research.

Past work on patient-specific image-guided needle procedures uses local control to compensate for errors induced by tissue deformation but does not preoperatively consider these effects [192]. Conversely, our sensorless planner searches for a globally optimal insertion plan but does not consider anomalies that may occur during execution. In the long run, we believe in combining these methods to create a pre-operative plan that is optimal under uncertainty and then use information from real-time imaging, when available, to correct deviations from the pre-operative plan.

4 Motion Planning in Deformable Soft Tissue with Obstacles with Applications to Needle Steering

In this chapter we introduce motion planning for *steerable needles*, a new class of needles that can follow curved paths around obstacles to reach clinical targets in soft tissue. Steerable needles are capable of reaching targets inaccessible by rigid needles.

We introduce a simulation and planner for steerable bevel-tip needle insertion that compensates for errors that occur due to tissue deformations. As in chapter 3, we begin with a simulation and planner in a 2-D imaging plane. Our interactive simulation approximates soft tissues as linearly elastic materials and uses a 2-D finite element model to compute tissue deformations due to tip and friction forces applied by the steerable needle. Polygonal obstacles represent tissues that cannot be cut by the needle, such as bone, or sensitive tissues that should not be damaged, such as nerves or arteries. The simulation enforces nonholonomic constraints on needle motion.

Our planner considers 4 degrees of freedom: initial location, initial orientation, binary bevel rotation, and insertion distance. The planner computes locally optimal values for these variables to compensate for tissue deformations and reach the target in simulation while avoiding polygonal obstacles and minimizing insertion distance so less tissue is damaged by the needle. Even in situations where real-time imaging such as ultrasound or interventional MRI is available, pre-planning is valuable to set the needle initial location and orientation and compute a desired trajectory that minimizes tissue damage.

4.1 Background on Needle Steering

Steerable needles follow curved paths when inserted into soft tissues. O'Leary et al. showed that needles with bevel tips bend more than symmetric-tip needles [170]. Webster et al. developed thin highly flexible bevel-tip needles using Nitinol and experimentally tested them in stiff tissue phantoms [208]. The needles followed constant-curvature paths in a plane when bevel rotation was fixed during needle insertion. Webster et al. [208] then developed a nonholonomic model for the steering flexible bevel-tip needles in rigid tissues. The nonholonomic model,

R. Alterovitz and K. Goldberg: Motion Planning in Medicine, STAR 50, pp. 45–55, 2008.
springerlink.com

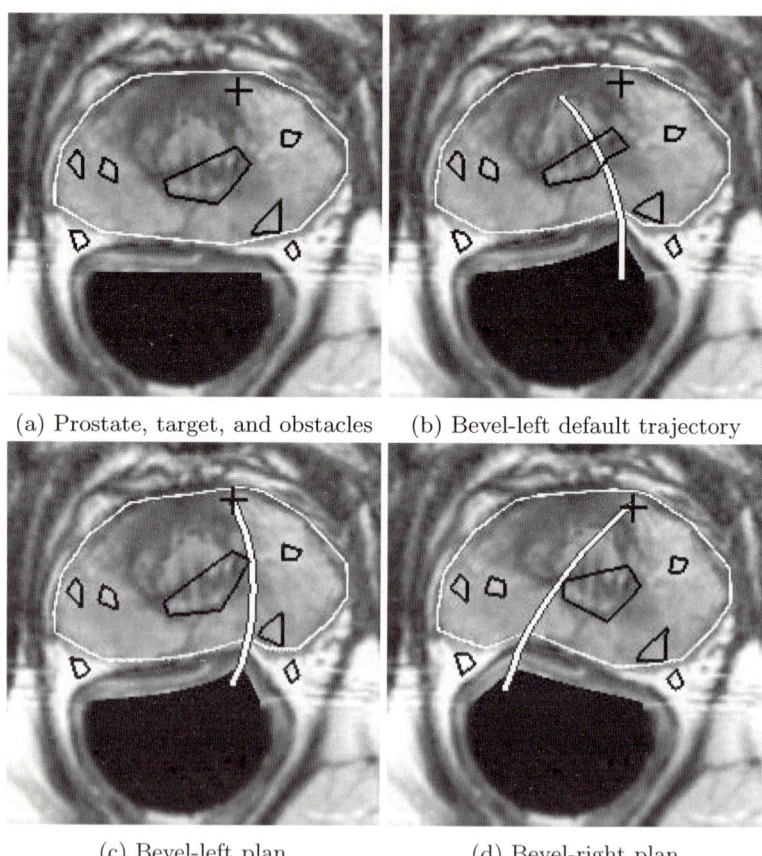

(a) Prostate, target, and obstacles   (b) Bevel-left default trajectory

(c) Bevel-left plan   (d) Bevel-right plan

Fig. 4.1. In this example based on an MR image of the prostate [121], a biopsy needle attached to a rigid rectal probe (black half-circle) is inserted into the prostate in simulation. Obstacles (polygons) and the target (cross) are overlaid on the image. The target is not accessible from the rigid probe by a straight line path without intersecting obstacles. However, bevel-tip needles bend as they are inserted into soft tissue (b). Our planner computes a locally optimal bevel-left needle insertion plan that reaches the target, avoids obstacles, and minimizes insertion distance (c). Using different initial conditions, our planner generates a plan for a bevel-right needle (d). Due to tissue deformation, the needle paths do not have constant curvature.

a generalization of a 3 degree-of-freedom bicycle model, was experimentally validated using a stiff tissue phantom.

Recent work has used a nonholonomic model of needle steering as a basis for motion planning in rigid tissues. Park et al. modeled 3-D needle steering using a unicycle model and used a diffusion-based method for planning without obstacles [171]. This work was based on advances by Zhou and Chirikjian in nonholonomic path planning, including stochastic model-based motion planning to compensate for noise bias [224] and probabilistic models of dead-reckoning error in nonholonomic robots [223]. A more recent method for needle steering uses a screw-based model and optimization to compute locally optimal paths that avoid spherical obstacles in 3-D in seconds of computation time [75].

Past work has addressed steering symmetric-tip needles in 2-D deformable tissue that have 3 degrees of freedom: translating the needle base perpendicular to the insertion direction, rotating the the needle base along an axis perpendicular to the plane of the tissue, and translation along the needle insertion axis [72, 92]. DiMaio and Salcudean compute and invert a Jacobian matrix to translate and orient the base to avoid point obstacles with oval-shaped potential fields. Glozman and Shoham approximate the tissue using springs and also use an inverse kinematics approach to translate and orient the base every time step. In our work, we address bevel-tip steerable needles that have 2 degrees of freedom during insertion: rotation about the insertion axis and translation along the insertion axis.

4.2   Simulating Needle Steering

A bevel-tip needle, unlike a symmetric-tip needle, will exert asymmetric forces on the surrounding soft tissue when it is inserted. This causes the needle to cut tissue at an angle, as shown in figure 4.2. If the needle is sufficiently flexible, the asymmetric forces and angled cutting will cause the needle to bend in the direction of the bevel.

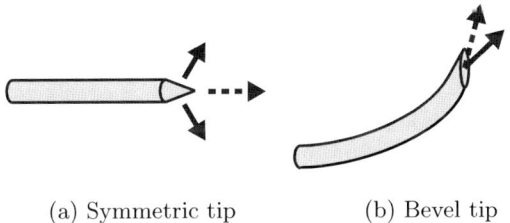

(a) Symmetric tip   (b) Bevel tip

Fig. 4.2. A symmetric-tip needle (a) exerts forces (solid vectors) on the tissue equally in all directions, so it cuts tissue (dashed vector) in the direction that the tip is moving. A bevel-tip needle (b) exerts forces asymmetrically and cuts tissue at an angle.

In 2-D, we only consider 2 bevel rotations: bevel-right (0°) and bevel-left (180°), as shown in figure 4.1. Rotating the bevel to different orientations will cause the needle tip to move out of the imaging plane. In future work, we plan to extend our 2-D model to 3-D and consider any bevel orientation in the range [0°, 360°).

Our simulation models forces exerted by the needle on the soft tissue, including the cutting force at the needle tip and friction forces along the needle shaft. We assume needle bending forces are negligible compared to the elastic forces applied by the soft tissue to the needle.

4.2.1   Soft Tissue Model

We specify the anatomy geometry (i.e. the prostate and surrounding tissues) using a finite element mesh. The geometric input is a 2-D slice of tissue with tissue types segmented by polygons. We automatically generate a finite element mesh G composed of n nodes and m triangular elements in a regular right triangle mesh or using the constrained Delaunay triangulation software program Triangle [191], which generates meshes that conform to the segmented tissue type polygons.

The model must also include tissue material properties and boundary conditions for the finite element mesh. In our current implementation, we approximate soft tissues as linearly elastic, homogeneous, isotropic materials. For each segmented tissue type, the model requires tissue material properties (Young's modulus, Poisson's ratio, and density). We set values for these parameters as described in past work [18]. Mesh nodes inside bones are constrained to be fixed. A boundary condition of either free or fixed must be specified for each node on the mesh perimeter.

The complete tissue model M specifies the finite element mesh G, material properties, and boundary conditions. We assume the tissue in M is initially at equilibrium and ignore external forces not applied by the needle. We do not model physiological changes such as edema (tissue swelling), periodic tissue motion due to breathing or heart beat, or slip between tissue type boundaries.

4.2.2   Computing Soft Tissue Deformations

The *material mesh* G defines the geometry of the undeformed tissues, with each node i having coordinate $\mathbf{x}_i$ in the material frame. Forces resulting from needle insertion cause the tissue to deform. The deformation is defined by a displacement $\mathbf{u}_i$ for each node i in mesh G. The *deformed mesh* G' is constructed in the world frame using the displaced node coordinate $\mathbf{x}_i + \mathbf{u}_i$ for each node i. A point $\mathbf{y}$ in the material frame can be transformed to the world frame coordinate $\mathbf{y}'$ and vice versa using linear interpolation between the nodes of the enclosing finite element [17].

At each time step of the simulation we compute the acceleration of each node i, which includes acceleration due to elastic forces computed using a linear finite element method and the external force $\mathbf{f}_i$ exerted by the needle. We use explicit Euler time integration to integrate velocity and displacement for each free node for each time step. Time steps have duration $h = 0.01$ seconds.

4.2.3   Needle Insertion Model

Without loss of generality, we set the coordinate axes of the world frame so that the default needle insertion axis is along the positive z-axis. The y-axis corresponds to the initial location degree of freedom. The needle tip is initially located at a base coordinate $\mathbf{p}_0 = (y_0, z_0)$. The initial orientation of the needle is specified using θ, as shown in figure 4.3. For simulation stability, we constrain θ between $-45°$ and $45°$. The needle tip rotation is either bevel-right ($0°$) or bevel-left ($180°$). We assume the needle tip rotation is held constant during insertion

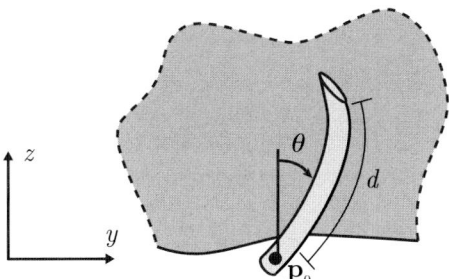

Fig. 4.3. Slice of soft tissue in the yz plane. The bevel-tip needle is initially at the base coordinate $\mathbf{p}_0$ with orientation θ. It is inserted a distance d, causing the surrounding soft tissue to deform.

due to planner efficiency and lack of experimental data for simulation, although we hope to relax this assumption in future work.

We assume the flexible needle is supported so that it does not bend outside the tissue. Once the needle has entered the tissue, it will bend in the direction of the bevel-tip. The distance the needle has been inserted from the base coordinate is d. We parameterize the needle by s where $s = 0$ corresponds the needle base and $s = d$ corresponds to the needle tip. Let $\mathbf{p}_s$ denote the material frame coordinate of the point along the needle a distance s from the base.

Simulation of needle insertion requires a needle model N that specifies needle properties, including insertion velocity v, the cutting force required at the needle tip to cut tissue, and the static and dynamic coefficients of friction between the tissue and needle.

We model the needle by line segment elements that correspond to edges of triangle elements in the deformed tissue mesh. Since the needle path is not known a priori, the material mesh must be modified in real-time. The simulation maintains a node at the needle tip location and a list of nodes along the needle shaft. At each simulation time step, the needle exerts force on the tissue at the needle tip, where the needle is displacing and cutting the tissue, and along the needle shaft due to friction.

Highly flexible bevel-tip needles tested in tissue phantoms by Webster et al. [208] were experimentally shown to follow a constant-curvature path when the bevel rotation was fixed during needle insertion. Setting simulation parameters to the limiting case of highly stiff tissue, zero tissue cutting force, and zero friction allows us to replicate this constant curvature path. In other cases, the needle path through deformed tissue may not be of constant curvature.

4.2.4   Simulating Cutting at the Needle Tip

During each simulation time step, the needle tip moves a distance vh in the world frame, where v is the needle insertion velocity and h is the time step duration. The simulation must maintain element edges along the needle path, which requires mesh modification as the needle cuts through the tissue.

The simulation constrains a node to be located at the needle tip. The current needle tip node is labeled n_{tip} and the needle is pointed in direction $\mathbf{q}$. The needle will cut tissue a small distance d_{cut} along the vector $\mathbf{r}$ in the world frame, where $\mathbf{r}$ is deflected from $\mathbf{q}$ by an angle θ_d, as shown in figure 4.4. If the force at the needle tip along $\mathbf{r}$ is greater than a threshold f_{cut} based on needle and tissue properties, then the needle will cut through the tissue. Cutting is represented in the material mesh by moving the needle tip node n_{tip} by the distance d_{cut} transformed to the material frame. If no tissue deformation occurs, this method guarantees the needle will cut a path of constant curvature whose radius of curvature is a function of the deflection angle θ_d. When tissue deformation does occur at the needle tip, the path will be of constant curvature locally but will deviate from constant curvature globally depending on the magnitude of the deformations.

As the needle tip cuts through the mesh, it will be necessary to change the needle tip node. If the needle tip node is too close to the opposite triangle edge e, the tip node is moved back along the shaft and a new tip node, the closest node along edge e, is selected as the new tip node and moved to the new tip location in the material frame.

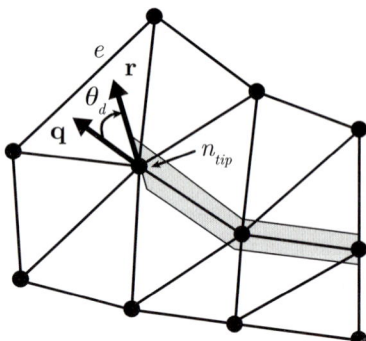

Fig. 4.4. The tissue mesh is modified so edge boundaries are formed along the path of needle insertion. A subset of the tissue mesh, centered at needle tip node n_{tip}, is shown. The straight line path of the needle is shown by vector $\mathbf{q}$. Because of the bevel-tip, the needle cuts tissue in direction $\mathbf{r}$, which is deflected from $\mathbf{q}$ by θ_d degrees.

4.2.5   Simulating Friction Along the Needle Shaft

We implemented a stick-slip friction model between the needle and the soft tissue. Nodes along the needle shaft carry friction state information; they are either attached to the needle (in the static friction state) or allowed to slide along the needle shaft (in the dynamic friction state).

When a node enters the static friction state, its distance from the needle tip along the shaft is computed. For each time step where the node remains in the static friction state, its position is modified by moving it tangent to the needle so that its distance from the tip along the needle shaft is held constant. A node moves from the static to the dynamic friction state when the force required to displace the node along the needle shaft exceeds a slip force parameter $f_{s_{max}}$.

When a node is in the dynamic friction state, a dissipative force is applied along the needle tangent. A node moves from the dynamic to the static friction state when the relative velocity of the needle to the tissue at the node is close to zero.

4.2.6 Simulation Results

Our simulator was implemented in C++ using OpenGL for visualization. It achieved an average of approximately 100 frames per second on a 1.6GHz Pentium M computer for a mesh composed of 1250 triangular elements. Computation time per frame increases linearly with the number of nodes along the needle shaft.

We demonstrate our simulation results in 2 cases: rigid tissue and deformable tissue. In both cases we simulate the insertion of a bevel-tip needle into a square of tissue fixed on 3 sides. In the first case, we consider tissue that is stiff relative to the needle and a sharp frictionless bevel-tip needle that cuts the tissue with zero cutting force. As shown in Fig 4.5(a), the simulated needle follows a path of constant curvature, which is the behavior experimentally verified by Webster et al. [208]. In the second case shown in figure 4.5(b), we insert the needle into a deformable soft tissue mesh with positive cutting force and friction coefficients.

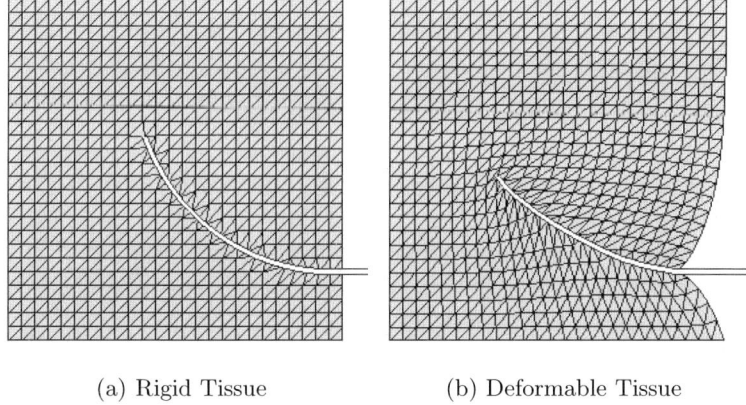

(a) Rigid Tissue   (b) Deformable Tissue

Fig. 4.5. We simulate insertion of a bevel-tip needle into a square tissue fixed on 3 sides. When the tissue is stiff relative to the needle, a sharp frictionless needle cuts a path of constant curvature (a). A needle with positive cutting and friction forces will bend in deformable tissue (b).

Although the tip locally follows a path of constant curvature as explained in Section 4.2.4, the global path is not of constant curvature. Past experiments have demonstrated the effect of tissue deformations due to rigid needle insertion [17, 70]. We plan to develop experiments to test the bending behavior of flexible bevel-tip needles in deformable tissues to more accurately set parameters for our model in future work.

4.3   Motion Planning for Needle Steering

A needle steering plan is defined by $X = (y_0, \theta, b, d)$ where $y_0 \in \mathbb{R}$ is the insertion location, $\theta \in [-90°, 90°]$ is the insertion angle, $b \in \{0°, 180°\}$ is the bevel rotation, and $d \in \mathbb{R}^+$ is the distance the needle will be inserted. Obstacles are defined as non-overlapping polygons in a set O. The target is defined as a point $\mathbf{t}$ in the material frame of the soft tissue mesh. A plan X is feasible if the needle tip is within $\epsilon_t > 0$ of the target and the needle path in deformable tissue does not intersect any obstacle. The goal of needle insertion planning is to generate a feasible plan X that minimizes d.

4.3.1   Problem Formulation

The simulation of needle insertion described in Section 4.2 takes parameters X for the initial conditions and needle insertion distance, M for the soft tissue model, and N for the needle model and outputs the coordinates $\mathbf{p}_s$ for $s \in [0, d]$, which the needle will follow in the material frame.

$$\mathbf{p}_s = \text{NeedleSim}(X, M, N), s \in [0, d]$$

The variables of plan X are constrained by application specific limits y_{min}, y_{max}, θ_{min}, θ_{max}, and d_{max}.

$$y_{min} \leq y_0 \leq y_{max}$$
$$\theta_{min} \leq \theta \leq \theta_{max}$$
$$0 \leq d \leq d_{max}$$

These constraints enforce the limits of the simulation, such as the angle requirements in Section 4.2.3. In the biopsy example in figure 4.1, d_{max} is the maximum length of the needle and $y_{max} - y_{min}$ defines the width of the rectal probe.

The needle tip coordinate $\mathbf{p}_d$ in a feasible solution must be within Euclidean distance ϵ_t of the target $\mathbf{t}$.

$$\|\mathbf{p}_d - \mathbf{t}\| \leq \epsilon_t$$

In the presence of a nonempty set of polygonal obstacles O, we require that the needle path in a feasible solution does not intersect an obstacle. Let c_s be the distance from $\mathbf{p}_s$ to the closest point on the closest obstacle $o \in O$ and let the sign of c_s be negative if $\mathbf{p}_s$ is inside obstacle o and positive otherwise. We require $c_s \geq \epsilon_o$ for some given tolerance $\epsilon_o \geq 0$ for all points s along the needle shaft. We formulate this constraint as

$$\int_0^d \max\{-c_s + \epsilon_o, 0\}ds \leq 0.$$

We can quickly compute this integral numerically using points sampled along the needle path.

We summarize the problem formulation for variable $X = (y_0, \theta, b, d)$ given target coordinate $\mathbf{t}$, polygonal obstacles O, tolerances ϵ_t and ϵ_o, tissue model parameters M, needle model parameters N, and variable limits y_{min}, y_{max}, θ_{min}, θ_{max}, and d_{max}.

$$\min f(X) = d$$
$$\text{Subject to:}$$
$$\|\mathbf{p}_d - \mathbf{t}\| \leq \epsilon_t$$
$$\int_0^d \max\{-c_s + \epsilon_o, 0\}ds \leq 0$$
$$y_{min} \leq y_0 \leq y_{max}$$
$$\theta_{min} \leq \theta \leq \theta_{max}$$
$$0 \leq d \leq d_{max}$$

The values of $\mathbf{p}_s$ for $s \in [0, d]$ are computed by executing the simulator NeedleSim(X, M, N). The obstacle distances c_s for $s \in [0, d]$ are computed using $\mathbf{p}_s$ and the set of obstacles O.

4.3.2   Optimization Method

To reduce the complexity of the optimization, we reduce the number of variables in X from 4 to 2. Given a plan X, we can find the optimal insertion distance d by executing the simulation to insertion distance d_{max} and identifying the point $\mathbf{p}_s$ along the needle path that minimizes the distance to the target $\mathbf{t}$. Hence, d does not need to be explicitly treated as a variable since its value is implied by the other variables in X. Furthermore, variable b in X is binary since it represents the bevel-right or bevel-left needle rotation state. We optimize X separately for the bevel-right and bevel-left states.

We solve for a locally optimal solution X^* using a penalty method. Penalty methods, originally developed in the 1950's and 1960's, solve a constrained nonlinear optimization problem by converting it to a series of unconstrained nonlinear optimization problems [31]. Given the constrained optimization problem $\min f(\mathbf{x})$ subject to $g(\mathbf{x}) \leq 0$, we can write the unconstrained problem $\min(f(\mathbf{x}) + \mu \max\{0, g(\mathbf{x})\}^2)$ for some large $\mu > 0$. Penalty methods generate a series of unconstrained optimization problems as $\mu \to \infty$. Each unconstrained optimization problem can be solved using Gradient Descent or variants of Newton's Method. For convex nonlinear problems, the method will generate points that converge arbitrarily close to the global optimal solution [31]. For nonconvex problems, the method can only converge to a local optimal solution.

For steerable needle insertion planning, we convert the target and obstacle constraints to penalty functions to define a new nonlinear nonconvex optimization problem.

$$\min \hat{f}(X) = d + \mu \left(\max\{\|\mathbf{p}_d - \mathbf{t}\| - \epsilon_t, 0\}\right)^2 +$$
$$\mu \left(\int_0^d \max\{-c_s + \epsilon_o, 0\}ds\right)^2$$

Subject to:
$$y_{min} \leq y_0 \leq y_{max}$$
$$\theta_{min} \leq \theta \leq \theta_{max}$$
$$0 \leq d \leq d_{max}$$

Evaluating the objective function $\hat{f}(X)$ requires executing the simulator NeedleSim(X, M, N) to compute the needle path $\mathbf{p}_s$ for $s \in [0, d]$ and the obstacles distances c_s. The remaining constraints are the limit constraints that are required for simulation stability and can never be violated.

We use Gradient Descent to find a local optimal solution to the unconstrained minimization problem $\min \hat{f}(X)$. The limit constraints are easily enforced at each iteration. We solve a sequence of 4 unconstrained problems, each with 10 Gradient Descent iterations. After each unconstrained problem has been solved, we multiply the penalty factor μ by 10. In future work, we plan to determine problem-specific termination criteria for the unconstrained optimization problems and for the penalty method.

The objective function $\hat{f}(X)$ cannot be directly differentiated since the simulator cannot be written as a closed form equation. For the Gradient Descent method, we numerically approximate the derivatives of the objective function with respect to the insertion location y_0 and orientation θ. We compute $d\hat{f}/dy_0$ by translating the needle path by Δy_0 and recomputing $\hat{f}$. Similarly, we compute $d\hat{f}/d\theta$ by rotating the needle path by $\Delta\theta$ about the insertion base coordinate $\mathbf{p}_0$ and recomputing $\hat{f}$. These approximations do not explicitly account for the different deformations that occur when y_0 or θ are modified but were sufficiently accurate for small Δy_0 and $\Delta\theta$ in our results described below.

4.3.3   Planner Results

We implemented the planner in C++ and used the simulation described in Section 4.2. Results for medical biopsy examples are shown in figure 4.1(c), figure 4.1(d), and figure 4.6(b). For each example, the tissue model mesh was composed of 1196 triangular elements and the planner required approximately 5 minutes of computation time on a Pentium M 1.6GHz computer. We set $\epsilon_t = \epsilon_o = 0.1$cm and the penalty method solution satisfied the constraints within a tolerance of 0.02.

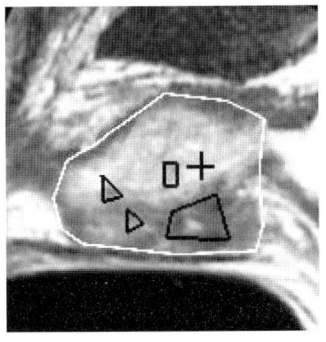
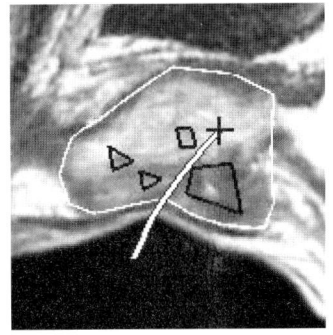

(a) Initial Conditions   (b) Bevel-right Plan

Fig. 4.6. In this example based on an MR image of the sagittal plane of the prostate [121], a biopsy needle is inserted into the prostate (a). The planner computes an initial position, orientation, and insertion distance so the needle reaches the target (cross) while avoiding obstacles (polygons) and compensating for tissue deformations in simulation (b).

4.4   Conclusion and Open Problems

We described a needle insertion planning algorithm for steerable bevel-tip needles that combines numerical optimization with soft tissue simulation. The simulation, based on a linear finite element method described in chapter 2, is a generalization of the simulation of rigid needles introduced in chapter 3. The simulation models the effects of steerable needle tip and frictional forces on soft tissues defined by a 2-D mesh. Our planning algorithm computes a locally optimal initial location, orientation, and insertion distance for the needle to compensate for predicted tissue deformations and reach a target while avoiding polygonal obstacles.

The effectiveness of the planner is dependent on the accuracy of the simulation of steerable needle insertion and soft tissue deformations. Future work that would improve on these results includes comparing the output of our simulation to new physical experiments, determining the sensitivity of results to model parameters, allowing bevel rotation during insertion, and extending the simulation and planner to 3-D.

5 Motion Planning for Curvature-Constrained Mobile Robots with Applications to Needle Steering

Advances in medical imaging modalities such as MRI, ultrasound, and x-ray fluoroscopy are now providing physicians with real-time, patient-specific information as they perform medical procedures such as extracting tissue samples for biopsies, injecting drugs for anesthesia, or implanting radioactive seeds for brachytherapy cancer treatment. These diagnostic and therapeutic medical procedures require insertion of a needle to a specific location in soft tissue. We are developing motion planning algorithms for medical needle insertion procedures that can utilize the information obtained by real-time imaging to accurately reach desired locations.

We consider a new class of medical needles, first introduced in chapter 4, that can be steered to targets in soft tissue that are inaccessible to traditional stiff needles. *Steerable needles* have two key properties: they are composed of a flexible material and have a bevel-tip. These properties enable steerable needles to follow curved paths through soft tissue. Steerable needles can be controlled by 2 degrees of freedom actuated at the needle base: insertion distance and bevel direction. Webster et al. experimentally demonstrated that, under ideal conditions, a flexible bevel-tip needle cuts a path of constant curvature in the direction of the bevel, and the needle shaft bends to follow the path cut by the bevel tip [208]. In a plane, a needle subject to this nonholonomic constraint based on bevel direction is equivalent to a Dubins car that can only steer its wheels far left or far right but cannot go straight.

The steerable needle motion planning problem is to determine a sequence of actions (insertions and direction changes) so the needle tip reaches the specified target while avoiding obstacles and staying inside the workspace. Given a segmented medical image of the target, obstacles, and starting location, the feasible workspace for motion planning is defined by the soft tissues through which the needle can be steered. Obstacles represent tissues that cannot be cut by the needle, such as bone, or sensitive tissues that should not be damaged, such as nerves or arteries.

In this chapter, we consider motion planning for steerable needles in the context of an image-guided procedure: real-time imaging and computer vision algorithms are used to track the position and orientation of the needle tip in the

R. Alterovitz and K. Goldberg: Motion Planning in Medicine, STAR 50, pp. 57–74, 2008.
springerlink.com

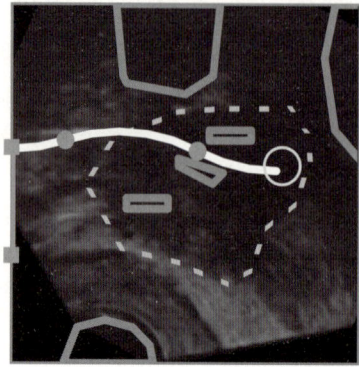
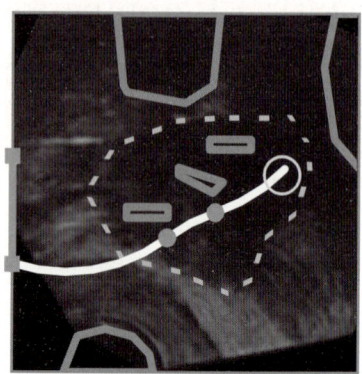

(a) Minimize path length
$p_s = 36.7\%$

(b) Maximize probability of success
$p_s = 73.7\%$

Fig. 5.1. Our motion planner computes actions (insertions and direction changes, indicated by dots) to steer the needle from an insertion entry region (vertical line on left between the solid squares) to the target (open circle) inside soft tissue, without touching critical areas indicated by polygonal obstacles in the imaging plane. The motion of the needle is not known with certainty; the needle tip may be deflected during insertion due to tissue inhomogeneities or other unpredictable soft tissue interactions. We explicitly consider this uncertainty to generate motion plans to maximize the probability of success, p_s, the probability that the needle will reach the target without colliding with an obstacle or exiting the workspace boundary. Relative to a planner that minimizes path length, our planner considering uncertainty may generate longer paths with greater clearance from obstacles to maximize p_s.

tissue. Recently developed methods can provide this information for a variety of imaging modalities [55, 67]. In this chapter, we consider motion plans in an imaging plane since the speed/resolution trade-off of 3-D imaging modalities is generally poor for 3-D real-time interventional applications. With imaging modalities continuing to improve, we will explore the natural extension of our planning approach to 3-D in future work.

5.0.1   Uncertainty and Motion Planning

Whereas many traditional motion planners assume a robot's motions are perfectly deterministic and predictable, a needle's motion through soft tissue cannot be predicted with certainty due to patient differences and the difficulty in predicting needle/tissue interaction. These sources of uncertainty may result in deflections of the needle's orientation, which is a type of slip in the motion of a Dubins car. Real-time imaging in the operating room can measure the needle's current position and orientation, but this measurement by itself provides no information about the effect of future deflections during insertion. Since the motion response of the needle is not deterministic, success of the procedure can rarely be guaranteed.

We develop a new motion planning approach for steering flexible needles through soft tissue that explicitly considers uncertainty in needle motion. To define optimality for a needle steering plan, we introduce a new objective for image-guided motion planning: *maximizing the probability of success*. In the case of needle steering, the needle insertion procedure continues until the needle reaches the target (success) or until failure occurs, where failure is defined as hitting an obstacle, exiting the feasible workspace, or reaching a state in which it is impossible to prevent the former two outcomes. Our method formulates the planning problem as a Markov Decision Process (MDP) based on an efficient discretization of the state space, models motion uncertainty using probability distributions, and computes optimal actions (within error due to discretization) for a set of feasible states using infinite horizon Dynamic Programming (DP).

Our motion planner is designed to run inside a feedback loop. After the feasible workspace, start region, and target are defined from a pre-procedure image, the motion planner is executed to compute the optimal action for each state. After the image-guided procedure begins, an image is acquired, the needle's current state (tip position and orientation) is extracted from the image, the motion planner (quickly) returns the optimal action to perform for that state, the action is executed and the needle may deflect due to motion uncertainty, and the cycle repeats.

In figure 5.1, we apply our motion planner in simulation to prostate brachytherapy, a medical procedure to treat prostate cancer in which physicians implant radioactive seeds at precise locations inside the prostate under ultrasound image guidance. In this ultrasound image of the prostate (segmented by a dotted line), obstacles correspond to bones, the rectum, the bladder, the urethra, and previously implanted seeds. Brachytherapy is currently performed in medical practice using rigid needles; here we consider steerable needles capable of obstacle avoidance. We compare the output of our new method, which explicitly considers motion uncertainty, to the output of a shortest-path planner that assumes the needles follow ideal deterministic motion. Our new method improves the expected probability of success by over 30% compared to shortest path planning, illustrating the importance of explicitly considering uncertainty in needle motion.

5.0.2   Background on Nonholonomic Motion Planning and MDP's

Nonholonomic motion planning has a long history in robotics and related fields [51, 136, 137, 139]. Past work has addressed deterministic curvature-constrained path planning where a mobile robot's path is, like a car, constrained by a minimum turning radius. Dubins showed that the optimal curvature-constrained trajectory in open space from a start pose to a target pose can be described using a discrete set of canonical trajectories composed of straight line segments and arcs of the minimum radius of curvature [74]. Jacobs and Canny considered polygonal obstacles and constructed a configuration space for a set of canonical trajectories [111], and Agarwal et al. developed a fast algorithm to compute a shortest path inside a convex polygon [4]. For Reeds-Shepp cars with reverse, Laumond et al. developed a nonholonomic planner using recursive subdivision of collision-free paths generated by a lower-level geometric planner [138], and Bicchi et al. proposed a technique that provides the shortest path for circular unicycles [36]. Sellen developed a discrete state-space approach; his discrete representation of orientation using a unit circle inspired our discretization approach [188].

Our planning problem considers steerable needles, a new type of needle currently being developed jointly by researchers at The Johns Hopkins University and The University of California, Berkeley [211]. Unlike traditional Dubins cars that are subject to a *minimum* turning radius, steerable needles are subject to a *constant* turning radius. Webster et al. showed experimentally that, under ideal conditions, steerable bevel-tip needles follow paths of constant curvature in the direction of the bevel tip [208], and that the radius of curvature of the needle path is not significantly affected by insertion velocity [209].

Park et al. formulated the planning problem for steerable bevel-tip needles in stiff tissue as a nonholonomic kinematics problem based on a 3-D extension of a unicycle model and used a diffusion-based motion planning algorithm to numerically compute a path [171]. The approach is based on recent advances by Zhou and Chirikjian in nonholonomic motion planning including stochastic model-based motion planning to compensate for noise bias [224] and probabilistic models of dead-reckoning error in nonholonomic robots [223]. Park's method searches for a feasible path in full 3-D space using continuous control, but it does not consider obstacle avoidance or the uncertainty of the response of the needle to insertion or direction changes, both of which are emphasized in our method.

In preliminary work on motion planning for bevel-tip steerable needles, we proposed an MDP formulation for 2-D needle steering [15] to find a stochastic shortest path from a start position to a target, subject to user-specified "cost" parameters for direction changes, insertion distance, and obstacle collisions. However, the formulation was not targeted at image-guided procedures, did not include insertion point optimization, and optimized an objective function that has no physical meaning. In this chapter, we develop a 2-D motion planning approach for image-guided needle steering that explicitly considers motion uncertainty to maximize the probability of success based on parameters that can be extracted from medical imaging without requiring user-specified "cost" parameters that may be difficult to determine.

MDP's and dynamic programming are ideally suited for medical planning problems because of the variance in characteristics between patients and the necessity for clinicians to make decisions at discrete time intervals based on limited known information. In the context of medical procedure planning, MDP's have been developed to assist in decisions such as timing for liver transplants [6], discharge times for severe sepsis cases [125], and start dates for HIV drug cocktail treatment [190]. MDP's and dynamic programming have also been used in a variety of robotics applications, including planning paths for mobile robots [63, 82, 139, 141].

Past work has investigated needle insertion planning in situations where soft tissue deformations are significant and can be modeled. Several groups have estimated tissue material properties and needle/tissue interaction parameters using tissue phantoms [61, 70] and animal experiments [100, 101, 117, 124, 168, 194]. Our past work addressed planning optimal insertion location and insertion distance for rigid symmetric-tip needles to compensate for 2-D tissue deformations predicted using a finite element model [17, 18, 19]. We previously also developed a different 2-D planner for bevel-tip steerable needles to explicitly compensate for the effects of tissue deformation by combining finite element simulation with numeric optimization [10]. This previous approach assumed that bevel direction can only be set once prior to insertion and employed local optimization that can fail to find a globally optimal solution in the presence of obstacles.

Past work has also considered insertion planning for needles and related devices capable of following curved paths through tissues using different mechanisms. One such approach uses slightly flexible symmetric-tip needles that are guided by translating and orienting the needle base to explicitly deform surrounding tissue, causing the needle to follow a curved path [71, 92]. DiMaio and Salcudean developed a planning approach that guides this type of needle around point obstacles with oval-shaped potential fields [71]. Glozman and Shoham also addressed symmetric-tip needles and approximated the tissue using springs [92]. Another steering approach utilizes a standard biopsy cannula (hollow tube needle) and adds steering capability with an embedded pre-bent stylet that is controlled by a hand-held, motorized device [169]. A recently developed "active cannula" device is composed of concentric, pre-curved tubes and is capable of following curved paths in a "snake-like" manner in soft tissue or open space [210].

Integrating motion planning for needle insertion with intra-operative medical imaging requires real-time localization of the needle in the images. Methods are available for this purpose for a variety of imaging modalities [55, 67]. X-ray fluoroscopy, a relatively low-cost imaging modality capable of obtaining images at regular discrete time intervals, is ideally suited for our application because it generates 2-D projection images from which the needle can be cleanly segmented [55].

Medical needle insertion procedures may also benefit from the more precise control of needle position and velocity made possible through robotic surgical assistants [103, 199]. Dedicated robotic hardware for needle insertion is being developed for a variety of medical applications, including stereotactic neurosurgery [151], CT-guided procedures [153], MR compatible surgical assistance [50, 68], thermotherapy cancer treatment [97], and prostate biopsy and therapeutic interventions [84, 186].

5.0.3   Overview of Motion Planning Method

In section 5.2, we first introduce a motion planner for Dubins cars with binary left/right steering subject to a *constant* turning radius rather than the typical *minimum* turning radius. This model applies to an idealized steerable needle whose motion is deterministic: the needle exactly follows arcs of constant curvature in response to insertion actions. Our planning method utilizes an efficient discretization of the state space for which error due to discretization can be tightly bounded. Since any feasible plan will succeed with 100% probability under the deterministic motion assumption, we apply the traditional motion planning objective of computing a shortest path plan from the current state to the target.

In section 5.3, we extend the deterministic motion planner to consider uncertainty in motion and introduce a new planning objective: maximize the probability of success. Unlike the objective function value of previous methods that consider motion uncertainty, the value of this new objective function has physical meaning: it is the probability that the needle tip will successfully reach the target during the insertion procedure. In addition to this intuitive meaning of the objective, our problem formulation has a secondary benefit: all data required for planning can be measured directly from imaging data without requiring tweaking of user-specified parameters. Rather than assigning costs to insertion distance, needle rotation, etc., which are difficult to estimate or quantify, our method only requires the probability distributions of the needle response to each feasible action, which can be estimated from previously obtained images.

Our method formulates the planning problem as a Markov Decision Process (MDP) and computes actions to maximize the probability of success using infinite horizon Dynamic Programming (DP). Solving the MDP using DP has key benefits particularly relevant for medical planning problems where feedback is provided at regular time intervals using medical imaging or other sensor modalities. Like a well-constructed navigation field, the DP solver provides an optimal action for any state in the workspace. We use the DP look-up table to automatically optimize the needle insertion point. Integrated with intra-operative medical imaging, this DP look-up table can also be used to optimally steer the needle in the operating room without requiring costly intra-operative re-planning. Hence, the planning solution can serve as a means of control when integrated with real-time medical imaging.

Throughout the description of the motion planning method, we focus on the needle steering application. However, the method is generally applicable to any car-like robot with binary left/right steering that follows paths composed of arcs of constant curvature, whose position can be estimated by sensors at regular intervals, and whose path may deflect due to motion uncertainty.

5.1 Problem Definition

Steerable bevel-tip needles are controlled by 2 degrees of freedom: insertion distance and rotation angle about the needle axis. The actuation is performed at the needle base outside the patient [208]. Insertion pushes the needle deeper into the tissue, while rotation turns the needle about its shaft, re-orienting the bevel at the needle tip. For a sufficiently flexible needle, Webster et al. experimentally demonstrated that rotating the needle base will change the bevel direction without changing the needle shaft's position in the tissue [208]. In the plane, the needle shaft can be rotated 180° about the insertion axis at the base so the

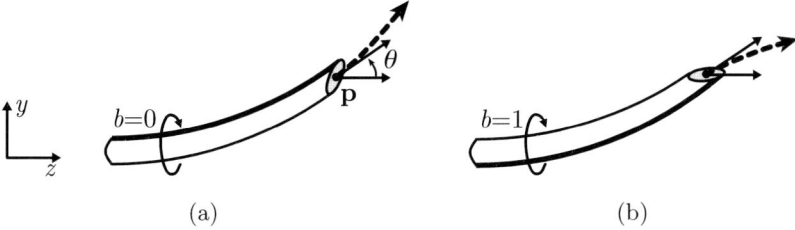

Fig. 5.2. The state of a steerable needle during insertion is characterized by tip position **p**, tip orientation angle θ, and bevel direction b (a). Rotating the needle about its base changes the bevel direction but does not affect needle position (b). The needle will cut soft tissue along an arc (dashed vector) based on bevel direction.

bevel points in either the bevel-left or bevel-right direction. When inserted, the asymmetric force applied by the bevel causes the needle to bend and follow a curved path through the tissue [208]. Under ideal conditions, the curve will have a constant radius of curvature r, which is a property of the needle and tissue. We assume the needle moves only in the imaging plane; a recently developed low-level controller using image feedback can effectively maintain this constraint [116]. We also assume the tissue is stiff relative to the needle and that the needle is thin, sharp, and low-friction so the tissue does not significantly deform. While the needle can be partially retracted and re-inserted, the needle's motion would be biased to follow the path in the tissue cut by the needle prior to retraction. Hence, in this chapter we only consider needle insertion, not retraction.

We define the workspace as a 2-D rectangle of depth z_{max} and height y_{max}. Obstacles in the workspace are defined by (possibly nonconvex) polygons. The obstacles can be expanded using a Minkowski sum with a circle to specify a minimum clearance [139]. The target region is defined by a circle with center point **t** and radius r_t.

As shown in figure 5.2, the state w of the needle during insertion is fully characterized by the needle tip's position $\mathbf{p} = (p_y, p_z)$, orientation angle θ, and bevel direction b, where b is either bevel-left ($b=0$) or bevel-right ($b=1$).

We assume the needle steering procedure is performed with image guidance; a medical image is acquired at regular time intervals and the state of the needle (tip position and orientation) is extracted from the images. Between image acquisitions, we assume the needle moves at constant velocity and is inserted a distance δ. In our model, direction changes can only occur at discrete *decision points* separated by the insertion distance δ. One of two actions u can be selected at any decision point: insert the needle a distance δ ($u = 0$), or change direction and insert a distance δ ($u = 1$).

During insertion, the needle tip orientation may be deflected by inhomogeneous tissue, small anatomical structures not visible in medical images, or local tissue displacements. Additional deflection may occur during direction changes due to stiffness along the needle shaft. Such deflections are due to an unknown aspect of the tissue structure or needle/tissue interaction, not errors in measurement of the needle's orientation, and can be considered a type of noise parameter in the plane. We model uncertainty in needle motion due to such deflections using probability distributions. The orientation angle θ may be deflected by some angle β, which we model as normally distributed with mean 0 and standard deviations σ_i for insertion ($u = 0$) and σ_r for direction changes followed by insertion ($u = 1$). Since σ_i and σ_r are properties of the needle and tissue, we plan in future work to automatically estimate these parameters by retrospectively analyzing images of needle insertion.

The goal of our motion planner is to compute an optimal action u for every feasible state w in the workspace to maximize the probability p_s that the needle will successfully reach the target.

5.2   Motion Planning for Deterministic Needle Steering

We first introduce a motion planner for an idealized steerable needle whose motion is deterministic: the needle perfectly follows arcs of constant curvature in response to insertion actions.

To computationally solve the motion planning problem, we transform the problem from a continuous state space to a discrete state space by approximating needle state $w = \{\mathbf{p}, \theta, b\}$ using a discrete representation. To make this approach tractable, we must round $\mathbf{p}$ and θ without generating an unwieldy number of states while simultaneously bounding error due to discretization.

5.2.1   State Space Discretization

Our discretization of the planar workspace is based on a grid of points with a spacing Δ horizontally and vertically. We approximate a point $\mathbf{p} = (p_y, p_z)$ by rounding to the nearest point $\mathbf{q} = (q_y, q_z)$ on the grid. For a rectangular workspace bounded by depth z_{max} and height y_{max}, this results in

$$N_s = \left\lfloor \frac{z_{max} + \Delta}{\Delta} \right\rfloor \left\lfloor \frac{y_{max} + \Delta}{\Delta} \right\rfloor$$

position states aligned at the origin.

Rather than directly approximating θ by rounding, which would incur a cumulative error with every transition, we take advantage of the discrete insertion distances δ. We define an *action circle* of radius r, the radius of curvature of the needle. Each point $\mathbf{c}$ on the action circle represents an orientation θ of the needle, where θ is the angle of the tangent of the circle at $\mathbf{c}$ with respect to the z-axis. The needle will trace an arc of length δ along the action circle in a counter-clockwise direction for $b = 0$ and in the clockwise direction for $b = 1$. Direction changes correspond to rotating the point $\mathbf{c}$ by 180° about the action circle origin and tracing subsequent insertions in the opposite direction, as shown in figure 5.3(a). Since the needle traces arcs of length δ, we divide the action circle into N_c arcs of length $\delta = 2\pi r/N_c$. The endpoints of the arcs generate a set of N_c action circle points, each representing a discrete orientation state, as

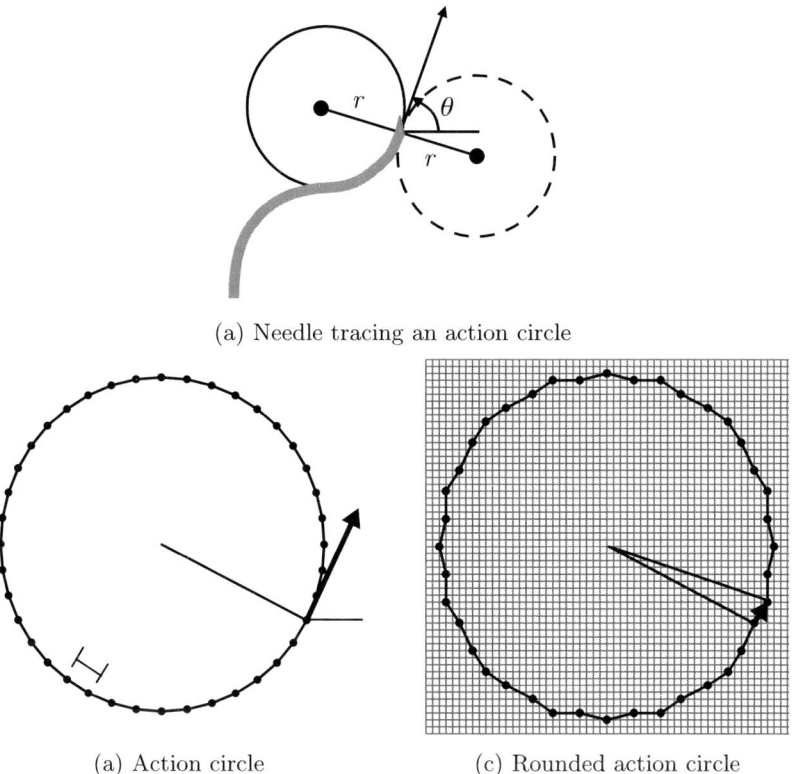

(a) Needle tracing an action circle

(a) Action circle   (c) Rounded action circle

Fig. 5.3. A needle in the bevel-left direction with orientation θ is tracing the solid action circle with radius r (a). A direction change would result in tracing the dotted circle. The action circle is divided into $N_c = 40$ discrete arcs of length δ (b). The action circle points are rounded to the nearest point on the Δ-density grid, and transitions for insertion of distance δ are defined by the vectors between rounded action circle points (c).

shown in figure 5.3(b). We require that N_c be a multiple of 4 to facilitate the orientation state change after a direction change.

At each of the N_s discrete position states on the Δ grid, the needle may be in any of the N_c orientation states and the bevel direction can be either $b = 0$ or $b = 1$. Hence, the total number of discrete states is $N = 2N_sN_c$.

Using this discretization, a needle state $w = \{\mathbf{p}, \theta, b\}$ can be approximated as a discrete state $s = \{\mathbf{q}, \Theta, b\}$, where $\mathbf{q} = (q_y, q_z)$ is the discrete point closest to $\mathbf{p}$ on the Δ-density grid and Θ is the integer index of the discrete action circle point with tangent angle closest to θ.

5.2.2 Deterministic State Transitions

For each state and action, we create a state transition that defines the motion of the needle when it is inserted a distance δ. We first consider the motion of the needle from a particular spatial state $\mathbf{q}$. To define transitions for each orientation state at $\mathbf{q}$, we overlay the action circle on a regular grid of spacing Δ and round the positions of the action circle points to the nearest grid point, as shown in figure 5.3(c). The displacement vectors between rounded action circle points encode the transitions of the needle tip. Given a particular orientation state Θ and bevel direction $b = 0$, we define the state transition using a translation component (the displacement vector between the positions of Θ and $\Theta - 1$ on the rounded action circle, which will point exactly to a new spatial state) and a new orientation state $(\Theta - 1)$. If $b = 1$, we increment rather than decrement Θ. We create these state transitions for each orientation state and bevel direction for each position state $\mathbf{q}$ in the workspace. This discretization of states and state transitions results in 0 discretization error in orientation when new actions are selected at δ intervals.

Certain states and transitions must be handled as special cases. States inside the target region and states inside obstacles are absorbing states. If the transition arc from a feasible state exits the workspace or intersects an edge of a polygonal obstacle, a transition to an obstacle state is used.

5.2.3 Discretization Error

Deterministic paths designated using this discrete representation of state will incur error due to discretization, but the error is bounded. At any decision point, the position error due to rounding to the Δ workspace grid is $E_0 = \Delta\sqrt{2}/2$. When the bevel direction is changed, a position error is also incurred because the distance between the center of the original action circle and the center of the action circle after the direction change will be in the range $2r \pm \Delta\sqrt{2}$. Hence, for a needle path with h direction changes, the final orientation is precise but the error in position is bounded above by $E_h = h\Delta\sqrt{2} + \Delta\sqrt{2}/2$.

5.2.4 Computing Deterministic Shortest Paths

For the planner that considers deterministic motion, we compute an action for each state such that the path length to the target is minimized. As in standard motion planning approaches [51, 136, 139], we formulate the motion planning problem as a graph problem. We represent each state as a node in a graph and state transitions as directed edges between the corresponding nodes. We merge all states in the target into a single "source" state. We then apply Dijkstra's shortest path algorithm [34] to compute the shortest path from each state to the target. The action u to perform at a state is implicitly computed based on the directed edge from that state that was selected for the shortest path.

5.3 Motion Planning for Needle Steering under Uncertainty

We extend the deterministic motion planner from section 5.2 to consider uncertainty in motion and to compute actions to explicitly maximize the probability of success p_s for each state. The planner retains the discrete approximation of the state space introduced in section 5.2.1, but replaces the single deterministic state transition per action defined in section 5.2.2 with a set of state transitions, each weighted by its probability of occurrence. We then generalize the shortest path algorithm defined in section 5.2.4 with a dynamic programming approach that enables the planner to utilize the probability-weighted state transitions to explicitly maximize the probability of success.

5.3.1   Modeling Motion Uncertainty

Due to motion uncertainty, actual needle paths will not always exactly trace the action circle introduced in section 5.2.1. The deflection angle β defined in section 5.1 must be approximated as discrete. We define discrete transitions from a state x_i, each separated by an angle of deflection of $\alpha = 360°/N_c$. In this chapter, we model β using a normal distribution with mean 0 and standard deviation σ_i or σ_r, and compute the probability for each discrete transition by integrating the corresponding area under the normal curve, as shown in figure 5.4. We set the number of discrete transitions N_{p_i} such that the areas on the left and right tails of the normal distribution sum to less than 1%. The left and right tail probabilities are added to the left-most and right-most transitions, respectively. Using this discretization, we define a transition probability matrix $P(u)$, where $P_{ij}(u)$ defines the probability of transitioning from state x_i to state x_j given that action u is performed.

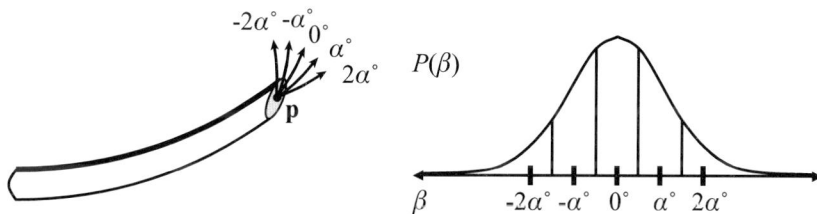

Fig. 5.4. When the needle is inserted, the insertion angle θ may be deflected by some angle β. We model the probability distribution of β using a normal distribution with mean 0 and standard deviation σ_i for insertion or σ_r for direction change. For a discrete sample of deflections ($\beta = \{-2\alpha, -\alpha, 0, \alpha, 2\alpha\}$), we obtain the probability of each deflection by integrating the corresponding area under the normal curve.

5.3.2   Maximizing the Probability of Success Using Dynamic Programming

The goal of our motion planning approach is to compute an optimal action u for every state w (in continuous space) such that the probability of reaching the target is maximized. We define $p_s(w)$ to be the probability of success given that the needle is currently in state w. If the position of state w is inside the target,

$p_s(w) = 1$. If the position of state w is inside an obstacle, $p_s(w) = 0$. Given an action u for some other state w, the probability of success will depend on the response of the needle to the action (the next state) and the probability of success at that next state. The expected probability of success is

$$p_s(w) = E[p_s(v)|w, u], \qquad (5.1)$$

where the expectation is over v, a random variable for the next state. The goal of motion planning is to compute an optimal action u for every state w:

$$p_s(w) = \max_u \left\{ E[p_s(v)|w, u] \right\}. \qquad (5.2)$$

For N discrete states, the motion planning problem is to determine the optimal action u_i for each state x_i, $i = 1, \ldots, N$. We re-write Eq. 5.2 using the discrete approximation and expand the expected value to a summation:

$$p_s(x_i) = \max_{u_i} \left\{ \sum_{j=1}^{N} P_{ij}(u_i) p_s(x_j) \right\}, \qquad (5.3)$$

where $P_{ij}(u_i)$ is the probability of entering state x_j after executing action u_i at current state x_i.

We observe that the needle steering motion planning problem is a type of MDP. In particular, Eq. 5.3 has the form of the Bellman equation for a stochastic maximum-reward problem [34]:

$$J^*(x_i) = \max_{u_i} \sum_{j=1}^{N} P_{ij}(u_i) \left(g(x_i, u_i, x_j) + J^*(x_j) \right). \qquad (5.4)$$

where $g(x_i, u_i, x_j)$ is a "reward" for transitioning from state x_i to x_j after performing action u_i. In our case, we set $J^*(x_i) = p_s(x_i)$, and we set $g(x_i, u_i, x_j) = 0$ for all x_i, u_i, and x_j. Stochastic maximum-reward problems of this form can be optimally solved using infinite horizon dynamic programming (DP).

Infinite horizon dynamic programming is a type of dynamic programming in which there is no finite time horizon [34]. For stationary problems, this implies that the optimal action at each state is purely a function of the state without explicit dependence on time. In the case of needle steering, once a state transition is made, the next action is computed based on the current position, orientation, and bevel direction without explicit dependence on past actions.

To solve the infinite horizon DP problem defined by the Bellman Eq. 5.4, we use the value iteration algorithm [34], which iteratively updates $p_s(x_i)$ for each state i by evaluating Eq. 5.3. This generates a DP look-up table containing the optimal action u_i and the probability of success $p_s(x_i)$ for $i = 1, \ldots, N$.

Termination of the algorithm is guaranteed in N iterations if the transition probability graph corresponding to some optimal stationary policy is acyclic [34]. Violation of this requirement will be rare in motion planning since a violation implies that an optimal action sequence results in a path that, with probability

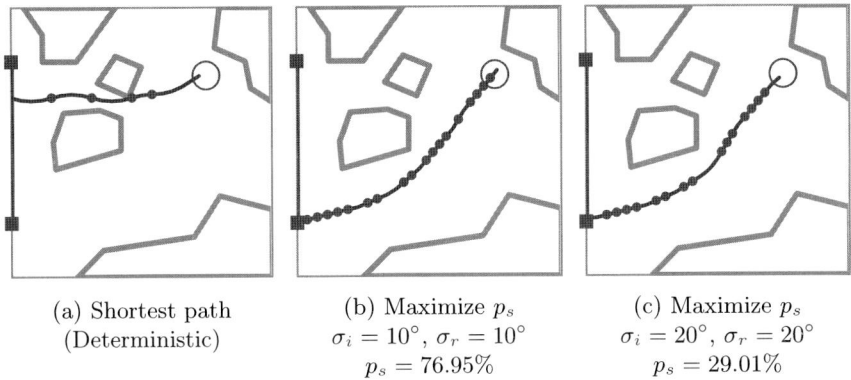

(a) Shortest path	(b) Maximize p_s	(c) Maximize p_s
(Deterministic)	$\sigma_i = 10°$, $\sigma_r = 10°$	$\sigma_i = 20°$, $\sigma_r = 20°$
	$p_s = 76.95\%$	$p_s = 29.01\%$

Fig. 5.5. As in figure 5.1, optimal plans maximizing the probability of success p_s illustrate the importance of considering uncertainty in needle motion. The shortest path plan passes through a narrow gap between obstacles (a). Since maximizing p_s explicitly considers uncertainty, the optimal expected path has greater clearance from obstacles, decreasing the probability that large deflections will cause failure to reach the target. Here we consider medium (b) and large (c) variance in tip deflections for a needle with smaller radius of curvature than in figure 5.1.

greater than 0, loops and passes through the same point at the same orientation more than once.

To improve performance, we take advantage of the sparsity of the matrices $P_{ij}(u)$ for $u = 0$ and $u = 1$. Each iteration of the value iteration algorithm requires matrix-vector multiplication using the transition probability matrix. Although $P_{ij}(u)$ has N^2 entries, each row of $P_{ij}(u)$ has only k nonzero entries, where $k << N$ since the needle will only transition to a state j in the spatial vicinity of state i. Hence, $P_{ij}(u)$ has at most kN nonzero entries. By only accessing nonzero entries of $P_{ij}(u)$ during computation, each iteration of the value iteration algorithm requires only $O(kN)$ rather than $O(N^2)$ time and memory. Thus, the total algorithm's complexity is $O(kN^2)$. To further improve performance, we terminate value iteration when the maximum change ϵ over all states is less than 10^{-3}, which in our test cases occurred in far fewer than N iterations, as described in section 5.4.

5.4 Computational Results

We implemented the motion planner in C++ and tested it on a 2.21GHz Athlon 64 PC. In figure 5.1, we set the needle radius of curvature $r = 5.0$, defined the workspace by $z_{max} = y_{max} = 10$, and used discretization parameters $N_c = 40$, $\Delta = 0.1$, and $\delta = 0.785$. The resulting DP problem contained $N = 800,000$ states. In all further examples, we set $r = 2.5$, $z_{max} = y_{max} = 10$, $N_c = 40$, $\Delta = 0.1$, and $\delta = 0.393$, resulting in $N = 800,000$ states.

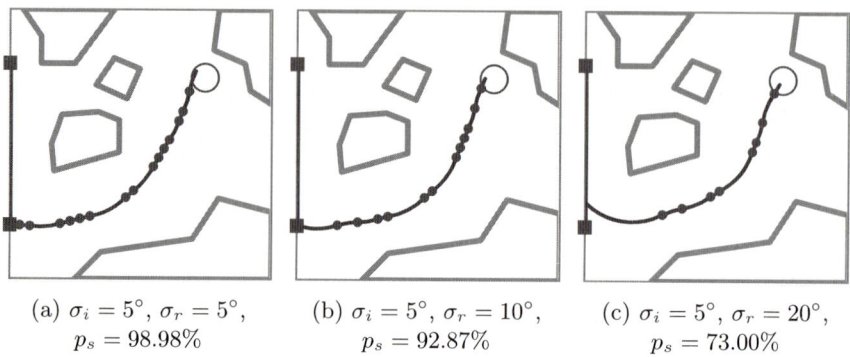

(a) $\sigma_i = 5°$, $\sigma_r = 5°$,
$p_s = 98.98\%$

(b) $\sigma_i = 5°$, $\sigma_r = 10°$,
$p_s = 92.87\%$

(c) $\sigma_i = 5°$, $\sigma_r = 20°$,
$p_s = 73.00\%$

Fig. 5.6. Optimal plans demonstrate the importance of considering uncertainty in needle motion, where σ_i and σ_r are the standard deviations of needle tip deflections that can occur during insertion and direction changes, respectively. For higher σ_r relative to σ_i, the optimal plan includes fewer direction changes. Needle motion uncertainty at locations of direction changes may be substantially higher than uncertainty during insertion due to transverse stiffness of the needle.

Optimal plans and probability of success p_s depend on the level of uncertainty in needle motion. As shown in Figs. 5.1 and 5.5, explicitly considering the variance of needle motion significantly affects the optimal plan relative to the shortest path plan generated under the assumption of deterministic motion. We also vary the variance during direction changes independently from the variance during insertions without direction changes. Optimal plans and probability of success p_s are highly sensitive to the level of uncertainty in needle motion due to direction changes. As shown in figure 5.6, the number of direction changes decreases as the variance during direction changes increases.

By examining the DP look-up table, we can optimize the initial insertion location, orientation, and bevel direction, as shown in Figs. 5.1, 5.5, and 5.6. In these examples, the set of feasible start states was defined as a subset of all states on the left edge of the workspace. By linearly scanning the computed probability of success for the start states in the DP look-up table, the method identifies the bevel direction b, insertion point (height y on the left edge of the workspace), and starting orientation angle θ (which varies from $-90°$ to $90°$) that maximizes probability of success, as shown in figure 5.7.

Since the planner approximates the state of the needle with a discrete state, the planner is subject to discretization errors as discussed in section 5.2.3. After each action, the state of the needle is obtained from medical imaging, reducing the discretization error in position of the current state to $\Delta\sqrt{2}/2$. However, when the planner considers future actions, discretization error for future bevel direction changes is cumulative. We illustrate the effect of cumulative discretization error during planning in figure 5.8, where the planner internally assumes the expected needle path will follow the dotted line rather than the actual expected

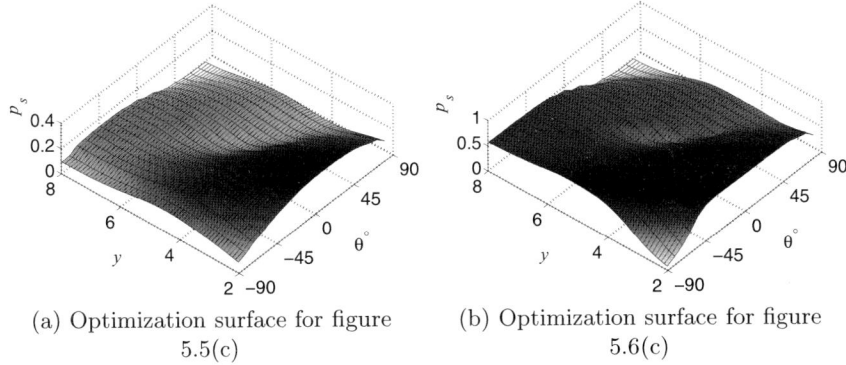

(a) Optimization surface for figure 5.5(c)

(b) Optimization surface for figure 5.6(c)

Fig. 5.7. The optimal needle insertion location y, angle θ, and bevel direction b are found by scanning the DP look-up table for the feasible start state with maximal p_s. Here we plot optimization surfaces for $b = 0$. The low regions correspond to states from which the needle has high probability of colliding with an obstacle or exiting the workspace, and the high regions correspond to better start states.

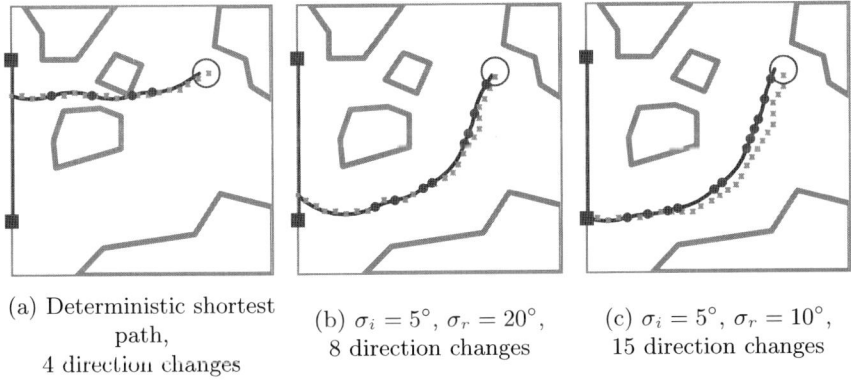

(a) Deterministic shortest path, 4 direction changes

(b) $\sigma_i = 5°$, $\sigma_r = 20°$, 8 direction changes

(c) $\sigma_i = 5°$, $\sigma_r = 10°$, 15 direction changes

Fig. 5.8. The small squares depict the discrete states used internally by the motion planning algorithm when predicting the expected path from the start state, while the solid line shows the actual expected needle path based on constant-curvature motion. The cumulative error due to discretization, which is bounded as described in section 5.2.3, is generally smaller when fewer direction changes (indicated by solid circles) are performed.

path indicated by the solid line. The effect of cumulative errors due to discretization, which is bounded as described in section 5.2.3, is generally smaller when fewer direction changes are planned.

As defined in section 5.3.2, the computational complexity of the motion planner is $O(kN^2)$. Fewer than 300 iterations were required for each example, with

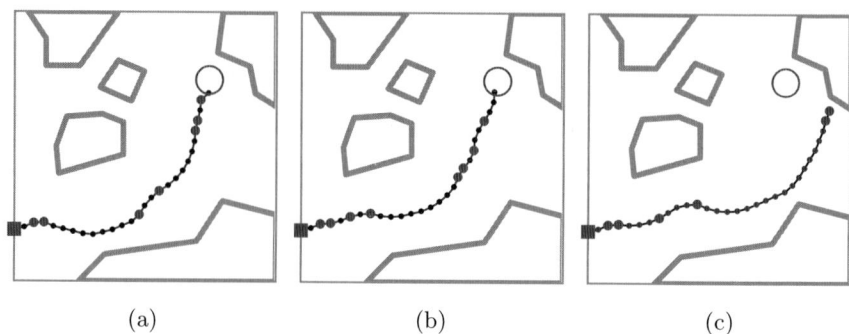

(a)   (b)   (c)

Fig. 5.9. Three simulated image-guided needle insertion procedures from a fixed starting point with needle motion uncertainty standard deviations of $\sigma_i = 5°$ during insertion and $\sigma_r = 20°$ during direction changes. After each insertion distance δ, we assume
the needle tip is localized in the image and identified using a dot. Based on the DP
look-up table, the needle is either inserted (small dots) or a direction change is made
(larger dots). The effect of uncertainty can be seen as deflections in the path, i.e.,
locations where the tangent of the path abruptly changes. Since $\sigma_r > \sigma_i$, deflections
are more likely to occur at points of direction change. In all cases, $p_s = 72.35\%$ at
the initial state. In (c), multiple deflections and the nonholonomic constraint on needle
motion prevent the needle from reaching the target.

fewer iterations required for smaller σ_i and σ_r. In all examples, the number of
transitions per state $k \leq 25$. Computation time to construct the MDP depends
on the collision detector used, as collision detection must be performed for all
N states and up to kN state transitions. Computation time to solve the MDP
for the examples ranged from 67 sec to 110 sec on a 2.21GHz AMD Athlon 64
PC, with higher computation times required for problems with greater variance,
due to the increased number of transitions from each state. As computation
only needs to be performed at the pre-procedure stage, we believe minutes of
computation time is reasonable for the intended applications. Intra-operative
computation time is effectively instantaneous since only a memory access to the
DP look-up table is required to retrieve the optimal action after the needle has
been localized in imaging.

Integrating intra-operative medical imaging with the pre-computed DP look-
up table could permit optimal steering of the needle in the operating room
without requiring costly intra-operative re-planning. We demonstrate the potential of this approach using simulation of needle deflections based on normal
distributions with mean 0 and standard deviations $\sigma_i = 5°$ and $\sigma_r = 20°$ in
figure 5.9. After each insertion distance δ, we assume the needle tip is localized
in the image. Based on the DP look-up table, the needle is either inserted or
the bevel direction is changed. The effect of uncertainty can be seen as deflections in the path, i.e., locations where the tangent of the path abruptly changes.
Since $\sigma_r > \sigma_i$, deflections are more likely to occur at points of direction change.

In practice, clinicians could monitor p_s, insertion length, and self-intersection while performing needle insertion.

5.5   Conclusion and Open Problems

We developed a new motion planning approach for steering flexible needles through soft tissue that explicitly considers uncertainty: the planner computes optimal actions to maximize the probability that the needle will reach the desired target. Motion planning for steerable needles, which can be controlled by 2 degrees of freedom at the needle base (bevel direction and insertion distance), is a variant of nonholonomic planning for a Dubins car with no reversals, binary left/right steering, and uncertainty in motion direction.

Given a medical image with segmented obstacles, target, and start region, our method formulates the planning problem as a Markov Decision Process (MDP) based on an efficient discretization of the state space, models motion uncertainty using probability distributions, and computes actions to maximize the probability of success using infinite horizon DP. Using our implementation of the method, we generated motion plans for steerable needles to reach targets inaccessible to stiff needles and illustrated the importance of considering uncertainty in needle motion, as shown in Figs. 5.1, 5.5, and 5.6.

Our approach has key features particularly beneficial for medical planning problems. First, the planning formulation only requires parameters that can be directly extracted from images (the variance of needle orientation after insertion with or without direction change). Second, we can determine the optimal needle insertion start pose by examining the pre-computed DP look-up table containing the optimal probability of success for each needle state, as demonstrated in figure 5.7. Third, intra-operative medical imaging can be combined with the pre-computed DP look-up table to permit optimal steering of the needle in the operating room without requiring time-consuming intra-operative re-planning, as shown in figure 5.9.

Extending this motion planner to 3-D would expand the applicability of the method. Although the mathematical formulation can be naturally extended, substantial effort will be required to geometrically specify 3-D state transitions and to efficiently handle the larger state space when solving the MDP. Extensions to 3-D should consider faster alternatives to the general value iteration algorithm, including hierarchical and adaptive resolution methods [27, 52, 158], methods that prioritize states [30, 63, 82, 95, 157], and other approaches that take advantage of the structure of our problem formulation [32, 40, 41].

Another open problem is to develop automated methods to estimate necessary parameters from medical images. These parameters include needle curvature and variance properties as well as the effects of including of multiple tissue types in the workspace with different needle/tissue interaction properties.

Our motion planner has implications beyond the needle steering application. We can directly extend the method to motion planning problems with a bounded number of discrete turning radii where current position and orientation can be measured but future motion response to actions is uncertain. For example, mobile robots subject to motion uncertainty with similar properties can receive periodic "imaging" updates from GPS or satellite images. Optimization of "insertion location" could apply to automated guided vehicles in a factory setting, where one machine is fixed but a second machine can be placed to maximize the probability that the vehicle will not collide with other objects on the factory floor. By identifying a relationship between needle steering and infinite horizon DP, we developed a motion planner capable of rigorously computing plans that are optimal in the presence of uncertainty.

6 The Stochastic Motion Roadmap: A Sampling-Based Framework for Planning with Motion Uncertainty

In many applications of motion planning, the motion of the robot in response to commanded actions cannot be precisely predicted. Whether maneuvering a vehicle over unfamiliar terrain, steering a flexible needle through human tissue to deliver medical treatment, guiding a micro-scale swimming robot through turbulent water, or displaying a folding pathway of a protein polypeptide chain, the underlying motions cannot be predicted with certainty. But in many of these cases, a probabilistic distribution of feasible outcomes in response to commanded actions can be experimentally measured. This stochastic information is fundamentally different from a deterministic motion model. Though planning shortest feasible paths to the goal may be appropriate for problems with deterministic motion, shortest paths may be highly sensitive to uncertainties: the robot may deviate from its expected trajectory when moving through narrow passageways in the configuration space, resulting in collisions.

In this chapter, we develop a new motion planning framework that explicitly considers uncertainty in robot motion at the planning stage. Because future configurations cannot be predicted with certainty, we define a plan by actions that are a function of the robot's current configuration. A plan execution is successful if the robot does not collide with any obstacles and reaches the goal. The idea is to compute plans that maximize the probability of success.

The approach we develop here is a generalization of the motion planning algorithm developed in chapter 5 to consider a wider range of robot motions and uncertainty models. Our framework builds on the highly successful approach used in Probabilistic Roadmaps (PRM's): a learning phase followed by a query phase [118]. During the learning phase, a random (or quasi-random) sample of discrete states is selected in the configuration space, and a roadmap is built that represents their collision-free connectivity. During the query phase, the user specifies initial and goal states, and the roadmap is used to find a feasible path that connects the initial state to the goal, possibly optimizing some criteria such as minimum length. PRM's have successfully solved many path planning problems for applications such as robotic manipulators and mobile robots [51, 139]. The term "probabilistic" in PRM comes from the random sampling of

R. Alterovitz and K. Goldberg: Motion Planning in Medicine, STAR 50, pp. 75–89, 2008.

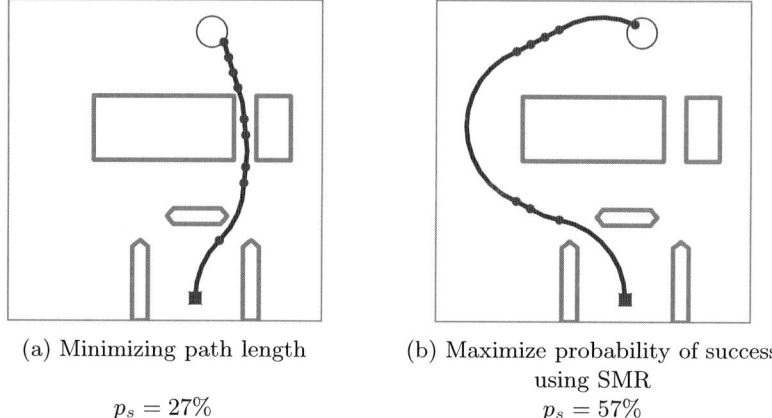

(a) Minimizing path length

$p_s = 27\%$

(b) Maximize probability of success
using SMR

$p_s = 57\%$

Fig. 6.1. The expected results of two plans to steer a Dubins-car mobile robot with left-right bang-bang steering and normally distributed motion uncertainty from an initial configuration (solid square) to a goal (open circle). Dots indicate steering direction changes. The Stochastic Motion Roadmap (SMR) introduces sampling of the configuration space and motion uncertainty model to generate plans that maximize the probability p_s that the robot will successfully reach the goal without colliding with an obstacle. Evaluation of p_s using multiple randomized simulations demonstrates that following a minimum length path under motion uncertainty (a) is substantially less likely to succeed than executing actions from an SMR plan (b).

states. An underlying assumption is that the collision-free connectivity of states is specified using boolean values rather than distributions.

In this chapter, we relax this assumption and combine a roadmap representation of the configuration space with a stochastic model of robot motion. The input to our method is a geometric description of the workspace and a motion model for the robot capable of generating samples of the next configuration that the robot may attain given the current configuration and an action. We require that the motion model satisfy the Markovian property: the distribution of the next state depends only on the action and current state, which encodes all necessary past history. As in PRM's, the method first samples the configuration space, where the sampling can be random [118], pseudo-random [140], or utility-guided [47]. We then sample the robot's motion model to build a *Stochastic Motion Roadmap (SMR)*, a set of weighted directed graphs with vertices as sampled states and edges encoding feasible state transitions and their associated probability of occurrence for each action.

The focus of our method is not to find a *feasible* motion plan, but rather to find an *optimal* plan that maximizes the probability that the robot will successfully reach a goal. Given a query specifying initial and goal configurations, we use the SMR to formulate a Markov Decision Process (MDP) where the "decision" corresponds to the action to be selected at each state in the roadmap. We solve

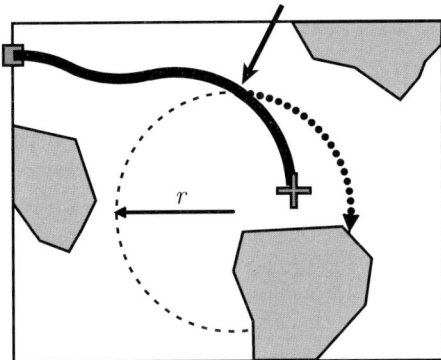

Fig. 6.2. The solid line path illustrates a motion plan from the start square to the goal (cross) for a nonholonomic mobile robot constrained to follow paths composed of continuously connected arcs of constant-magnitude curvature with radius of curvature r. If a deflection occurs at the location of the arrow, then the robot is unable to reach the goal due to the nonholonomic constraint, even if this deflection is immediately detected, since the robot cannot follow a path with smaller radius of curvature than the dotted line.

the MDP in polynomial time using Infinite Horizon Dynamic Programming. Because the roadmap is a discrete representation of the continuous configuration space and transition probabilities, the computed optimal actions are approximations of the optimal actions in continuous space that converge as the roadmap increases in size. Although the plan, defined by the computed actions, is fixed, the path followed by the robot may differ each time the plan is executed because different state transitions may occur due to motion uncertainty. As shown in figure 6.1, plans that explicitly consider uncertainty to maximize the probability of success can differ substantially from traditional shortest path plans.

In SMR, "stochastic" refers to the motion of the robot, not to the sampling of states. PRM's were previously extended to explore stochastic motion in molecular conformation spaces [21, 22], but without a planning component to optimize actions. Our SMR formulation is applicable to a variety of decision-based robotics problems. It is particularly suited for nonholonomic robots, for which a deflection in the path due to motion uncertainty can result in failure to reach the goal, even if the deflection does not result in an immediate collision. The expansion of obstacles using their Minkowski sum [139] with a circle corresponding to an uncertainty tolerance is often sufficient for holonomic robots for which deflections can be immediately corrected, but this does not address collisions resulting from a nonholonomic constraint as in figure 6.2. By explicitly considering motion uncertainty in the planning phase, we hope to minimize such failures.

Although we use the terms robot and workspace, SMR's are applicable to any motion planning problem that can be modeled using a continuous configuration space and discrete action set with uncertain transitions between configurations. In this chapter, we demonstrate a SMR planner using a variant of a Dubins car with bang-bang control, a nonholonomic mobile robot that can steer its wheels far left or far right while moving forward but cannot go straight. This model can generate motion plans for steerable needles, a new class of flexible bevel-tip medical needles introduced in chapter 4 that clinicians can steer through soft tissue around obstacles to reach targets inaccessible to traditional stiff needles [15, 208]. As in many medical applications, considering uncertainty is crucial to the success of medical needle insertion procedures: the needle tip may deflect from the expected path due to tissue inhomogeneities that cannot be detected prior to the procedure. Due to uncertainty in predicted needle/tissue interactions, needle steering is ill-suited to shortest-path plans that may guide the needle through narrow passageways between critical tissue such as blood vessels or nerves. By explicitly considering motion uncertainty using an SMR, we obtain solutions that result in possibly longer paths but that improve the probability of success.

6.0.1   Related Work

Motion planning can consider uncertainty in *sensing* (the current state of the robot and workspace is not known with certainty) and *predictability* (the future state of the robot and workspace cannot be deterministically predicted even when the current state and future actions are known) [139]. Extensive work has explored uncertainty associated with robot sensing, including Simultaneous Localization and Mapping (SLAM) and Partially Observable Markov Decision Processes (POMDP's) to represent uncertainty in the current state [51, 202]. In this chapter, we assume the current state is known (or can be precisely determined from sensors), and we focus on the latter type of uncertainty, predictability.

Predictability can be affected by uncertainty in the workspace and by uncertainty in the robot's motion. Previous and ongoing work addresses many aspects of uncertainty in the workspace, including uncertainty in the goal location, such as in pursuit-evasion games [139, 142], and in dynamic environments with moving obstacles [147, 204, 206]. A recently developed method for grasp planning uses POMDP's to consider uncertainty in the configuration of the robot and the state of the objects in the world [105]. In this chapter we focus explicitly on the case of uncertainty in the robot's motion rather than in goal or obstacle locations.

Apaydin et al. previously explored the connection between probabilistic roadmaps and stochastic motions using Stochastic Roadmap Simulation (SRS), a method designed specifically for molecular conformation spaces [21, 22]. SRS, which formalizes random walks in a roadmap as a Markov Chain, has been successfully applied to predict average behavior of molecular motion pathways of proteins and requires orders of magnitude less computation time than traditional Monte-Carlo molecular simulations. However, SRS cannot be applied to more general robotics problems, including needle steering, because the probabilities associated with state transitions are specific to molecular scale motions and the method does not include a planning component to optimize actions.

Considering uncertainty in the robot's response to actions during planning results in a stochastic optimal control problem where feedback is generally required for success. Motion planners using grid-based numerical methods and geometric analysis have been applied to robots with motion uncertainty (sometimes combined with sensing uncertainty) using cost-based objectives and worst-case analysis [39, 141, 143]. MDP's, a general approach that requires explicitly defining transition probabilities between states, have also been applied to motion planning by subdividing the workspace using regular grids and defining transition probabilities for motions between the grid cells [63, 81, 139]. This was the motion planning approach taken in chapter 5. These methods differ from SMR's since they use grids or problem-specific discretization.

Many existing planners for deterministic motion specialize in finding feasible paths through narrow passageways in complex configuration spaces using specialized sampling [20, 38] or learning approaches [158]. Since a narrow passageway is unlikely to be robust to motion uncertainty, finding these passageways is not the ultimate goal of our method. Our method builds a roadmap that samples the configuration space with the intent of capturing the uncertain motion transition probabilities necessary to compute optimal actions.

We apply SMR's to needle steering, a type of nonholonomic control-based motion planning problem. Nonholonomic motion planning has a long history in robotics and related fields [51, 139]. Past work has addressed deterministic curvature-constrained path planning with obstacles where a mobile robot's path is, like a car, constrained by a minimum turning radius [36, 111, 138, 152, 188]. For steerable needles, Park et al. applied a numeric diffusion-based method but did not consider obstacles or motion uncertainty [171]. Alterovitz et al. proposed an MDP formulation to find a stochastic shortest path for a steerable needle to a goal configuration, subject to user-specified "cost" parameters for direction changes, insertion distance, and collisions [15]. Because these costs are difficult to quantify, Alterovitz et al. introduced the objective function of maximizing probability of success [7]. These methods use a regular grid of states and an ad-hoc, identical discretization of the motion uncertainty distribution at all states. The methods do not consider the advantages of sampling states nor the use of sampling to estimate motion models.

6.0.2   SMR Contributions

SMR planning is a general framework that combines a roadmap representation of configuration space with the theory of MDP's to explicitly consider motion uncertainty at the planning stage to maximize the probability of success. SMR's use sampling to both learn the configuration space (represented as states) and to learn the stochastic motion model (represented as state transition probabilities). Sampling reduces the need for problem-specific geometric analysis or discretization for planning. As demonstrated by the success of PRM's, sampling states is useful for modeling complex configuration spaces that cannot be easily represented geometrically and extends well to higher dimensions. Random or quasi-random sampling reduces problems associated with regular grids of states, including the high computational complexity in higher dimensions and the sensitivity of solutions and runtimes to the selection of axes [139]. Sampling the stochastic motion model enables the use of a wide variety of motion uncertainty representations, including directly sampling experimentally measured data or using parameterized distributions such as a Gaussian distribution. This greatly improves previous Markov motion planning approaches that impose an ad-hoc, identical discretization of the transition probability distributions at all states (as was done in chapter 5).

Although SMR is a general framework, it provides improvements for steerable needle planning compared to previously developed approaches specifically designed for this application. Previous planners do not consider uncertainty in needle motion [171], or apply simplified models that only consider deflections at decision points and assume that all other motion model parameters are constant (as was done in chapter 5). Because we use sampling to approximate the motion model rather than a problem-specific geometric approximation, we eliminate the discretization error at the initial configuration and can easily include a more complex uncertainty model that considers arbitrary stochastic models for both insertion distance and radius of curvature. SMR's increase flexibility and decrease computation time; for problems with equal workspace size and expected values of motion model parameters, a query that requires over a minute to solve using a grid-based MDP due to the large number of states needed to bound discretization error (as in chapter 5) requires just 6 seconds using an SMR.

6.1   Algorithm

6.1.1   Input

To build an SMR, the user must first provide input parameters and function implementations to describe the configuration space and robot motion model. A configuration of the robot and workspace is defined by a vector $x \in C = \Re^d$, where d is the number of degrees of freedom in the configuration space C. At any configuration x, the robot can perform an action from a discrete action set U of size w. The bounds of the configuration space are defined by B_i^{min} and B_i^{max} for $i = 1, \ldots, d$, which specify the minimum and maximum values, respectively, for each configuration degree of freedom i. The functions isCollisionFree(x) and isCollisionFreePath(x, y) implicitly define obstacles within the workspace; the former returns false if configuration x collides with an obstacle and true otherwise, and the latter returns false if the path (computed by a local planner [118]) from configuration x to y collides with an obstacle and true otherwise. (We consider exiting the workspace as equivalent to colliding with an obstacle.) The function distance(x, y) specifies the distance between two configurations x and y, which can equal the Euclidean distance in d-dimensional space or some other user-specified distance metric. The function generateSampleTransition(x, u) implicitly defines the motion model and its probabilistic nature; this function returns a sample from a known probability distribution for the next configuration given that the robot is currently in configuration x and will perform action u.

Algorithm   81

Procedure 1. `buildSMR`

Require:
n: number of nodes to place in the roadmap
U: set of discrete robot actions
m: number of sample points to generate for each transition

Ensure:
SMR containing states V and transition probabilities (weighted edges) E^u for each action $u \in U$

  $V \leftarrow \emptyset$
  for all $u \in U$ **do**
    $E^u \leftarrow \emptyset$
  end for
  while $|V| < n$ **do**
    $q \leftarrow$ random state sampled from the configuration space
    if `isCollisionFree`(q) **then**
      $V \leftarrow V \cup \{q\}$
    end if
  end while
  for all $s \in V$ **do**
    for all $u \in U$ **do**
      for all $(t, p) \in$ `getTransitions`(V, s, u, m) **do**
        $E^u \leftarrow E^u \cup \{(s, t, p)\}$
      end for
    end for
  end for
  return weighted directed graphs $G^u = (V, E^u) \; \forall \, u \in U$

6.1.2 Building the Roadmap

We build the stochastic motion roadmap using the algorithm `buildSMR` defined in Procedure 1. The roadmap is defined by a set of vertices V and sets of edges E^u for each action $u \in U$. The algorithm first samples n collision-free states in the configuration space and stores them in V. In our implementation, we use a uniform random sampling of the configuration space inside the bounds defined by (B_i^{min}, B_i^{max}) for $i = 1, \ldots, d$, although other random distributions or quasi- random sampling methods could be used [51, 140]. For each state $s \in V$ and an action $u \in U$, `buildSMR` calls the function `getTransitions`, defined in Procedure 2, to obtain a set of possible next states in V and probabilities of entering those states when action u is performed. We use this set to add to E^u weighted directed edges (s, t, p), which specify the probability p that the robot will transition from state $s \in V$ to state $t \in V$ when currently in state s and executing action u.

The function `getTransitions`, defined in Procedure 2, estimates state transition probabilities. Given the current state s and an action u, it calls the problem-specific function `generateSampleTransition`(x, u) to generate a sample configuration q and then selects the state $t \in V$ closest to q using the problem-specific `distance` function. We repeat this motion sampling m times and then

---

Procedure 2. `getTransitions`

Require:
  V: configuration space samples
  s: current robot state, $s \in V$
  u: action that the robot will execute, $u \in U$
  m: number of sample points to generate for this transition
Ensure:
  List of tuples (t, p) where p is the probability of transitioning from state $s \in V$ to state $t \in V$ after executing u.

$R \leftarrow \emptyset$
for $i = 1$ to m **do**
  $q = \texttt{generateSampleTransition}(\texttt{s}, \texttt{u})$
  if $\texttt{isCollisionFreePath}(s, q)$ **then**
    $t \leftarrow \arg\min_{t \in V} \texttt{distance}(q, t)$
  else
    $t \leftarrow$ obstacle state
  end if
  if $(t, p) \in R$ for some p **then**
    Remove (t, p) from R
    $R \leftarrow R \cup \{(t, p + 1/m)\}$
  else
    $R \leftarrow R \cup \{(t, 1/m)\}$
  end if
end for
return R

---

estimate the probability of transitioning from state s to t as the proportion of times that this transition occurred out of the m samples. If there is a collision in the transition from state s to t, then the transition is replaced with a transition from s to a dedicated "obstacle state," which is required to estimate the probability that the robot collides with an obstacle.

This algorithm has the useful property that the transition probability from state s to state t in the roadmap equals the fraction of transition samples that fall inside state t's Voronoi cell. This property is implied by the use of nearest neighbor checking in `getTransitions`. As $m \to \infty$, the probability p of transitioning from s to t will approach, with probability 1, the integral of the true transition distribution over the Voronoi cell of t. As the number of states $n \to \infty$, the expected volume V_t of the Voronoi cell for state t equals $V/n \to 0$, where V is the volume of the configuration space. Hence, the error in the approximation of the probability p due to the use of a discrete roadmap will decrease as n and m increase.

6.1.3   Solving a Query

We define a query by an initial configuration s^* and a set of goal configurations T^*.

Using the SMR and the query input, we build an $n \times n$ transition probability matrix $P(u)$ for each $u \in U$. For each tuple $(s, t, p) \in E^u$, we set $P_{st}(u) = p$ so

Algorithm   83

$P_{st}(u)$ equals the probability of transitioning from state s to state t given that action u is performed. We store each matrix $P(u)$ as a sparse matrix that only includes pointers to a list of non-zero elements in each row and assume all other entries are 0.

We define $p_s(i)$ to be the probability of success given that the robot is currently in state i. If the position of state i is inside the goal, $p_s(i) = 1$. If the position of state i is inside an obstacle, $p_s(i) = 0$. Given an action u_i for some other state i, the probability of success will depend on the response of the robot to the action and the probability of success from the next state. The goal of our motion planner is to compute an optimal action u_i to maximize the expected probability of success at every state i:

$$p_s(i) = \max_{u_i} \left\{ E[p_s(j)|i, u_i] \right\}, \tag{6.1}$$

where the expectation is over j, a random variable for the next state. Since the roadmap is a discrete approximation of the continuous configuration space, we expand the expected value in Eq. 6.1 to a summation:

$$p_s(i) = \max_{u_i} \left\{ \sum_{j \in V} P_{ij}(u_i) p_s(j) \right\}. \tag{6.2}$$

We observe that Eq. 6.2 is an MDP and has the form of the Bellman equation for a stochastic shortest path problem [34]:

$$J^*(i) = \max_{u_i} \sum_{j \in V} P_{ij}(u_i) \left(g(i, u_i, j) + J^*(j) \right). \tag{6.3}$$

where $g(i, u_i, j)$ is a "reward" for transitioning from state i to j after action u_i. In our case, $g(i, u_i, j) = 0$ for all i, u_i, and j, and $J^*(i) = p_s(i)$.

Stochastic shortest path problems of the form in Eq. 6.3 can be optimally solved using infinite horizon dynamic programming. For stationary Markovian problems, the configuration space does not change over time, which implies that the optimal action at each state is purely a function of the state without explicit dependence on time. Infinite horizon dynamic programming is a type of dynamic programming (DP) in which there is no finite time horizon [34]. Specifically, we use the value iteration algorithm [34], which iteratively updates $p_s(i)$ for each state i by evaluating Eq. 6.3. This generates a DP look-up table containing the optimal action u_i and the probability of success $p_s(i)$ for each $i \in V$.

The algorithm is guaranteed to terminate in n (the number of states) iterations if the transition probability graph corresponding to some optimal stationary policy is acyclic [34]. Violation of this requirement can occur in rare cases in which a cycle is feasible and deviating from the cycle will result in imminent failure. To remove this possibility, we introduce a small penalty γ for each transition by setting $g(i, u_i, j) = -\gamma$ in Eq. 6.3. Increasing γ has the effect of giving preference to shorter paths at the expense of a less precise estimate of the probability of success, where the magnitude of the error is (weakly) bounded by γn.

6.1.4 Computational Complexity

Building an SMR requires $O(n)$ time to create the states V, not including collision detection. Generating the edges in E^u requires $O(wn)$ calls to `generateTransitions`, where $w = |U|$. For computational efficiency, it is not necessary to consolidate multiple tuples with the same next state t; the addition $p + 1/m$ can be computed automatically during value iteration. Hence, each call requires $O(mdn)$ time using brute-force nearest neighbor checking. For certain low-dimensional configuration spaces, this can be reduced to $O\left(m\exp\left(d\right)\log\left(n\right)\right)$ using kd-trees [219]. Hence, the total time complexity of building an SMR is $O\left(wmdn^2\right)$ or $O\left(wm\exp\left(d\right)n\log\left(n\right)\right)$. This does not include the cost of n state collision checks and nm checks of collision free paths, which are problem-specific and may increase the computational complexity depending on the workspace definition.

Solving a query requires building the transition probability matrices and executing value iteration. Although the matrices $P_{ij}(u)$ each have n^2 entries, we do not store the zero entries as described above. Since the robot will generally only transition to a state j in the spatial vicinity of state i, each row of $P_{ij}(u)$ has only k nonzero entries, where $k << n$. Building the sparse matrices requires $O(wkn)$ time. By only accessing the nonzero entries of $P_{ij}(u)$ during the value iteration algorithm, each iteration for solving a query requires only $O(wkn)$ rather than $O(wn^2)$ time. Thus, the value iteration algorithm's total time complexity is $O(wkn^2)$ since the number of iterations is bounded by n. To further improve performance, we terminate value iteration when the maximum change ϵ over all states is less than some user-specified threshold ϵ^*. In our test cases, we used $\epsilon^* = 10^{-7}$, which resulted in far fewer than n iterations.

6.2 SMR for Medical Needle Steering

We assume the workspace is extracted from a medical image, where obstacles represent tissues that cannot be cut by the needle, such as bone, or sensitive tissues that should not be damaged, such as nerves or arteries. As in chapter 5, we consider motion plans in an imaging plane since the speed/resolution trade-off of 3-D imaging modalities is generally poor for 3-D interventional applications. We assume the needle follows paths of radius of curvature r and moves a distance δ between image acquisitions that are used to determine the current needle position and orientation. We do not consider motion by the needle out of the imaging plane or needle retraction, which modifies the tissue and can influence future insertions. When restricted to motion in a plane, the bevel direction can be set to point left ($b = 0$) or right ($b = 1$) [208]. Due to the nonholonomic constraint imposed by the bevel, the motion of the needle tip can be modeled as a bang-bang steering car, a variant of a Dubins car that can only turn its wheels far left or far right while moving forward [7, 208].

Clinicians performing medical needle insertion procedures must consider uncertainty in the needle's motion through tissue due to patient differences and the difficulty in predicting needle/tissue interaction. Bevel direction changes further

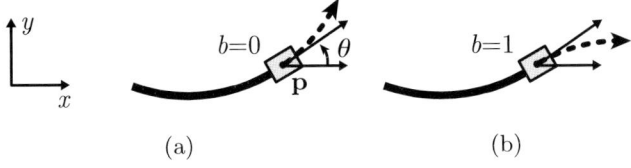

(a)   (b)

Fig. 6.3. The state of a bang-bang steering car is defined by point $\mathbf{p}$, orientation θ, and turning direction b (a). The car moves forward along an arc of constant curvature and can turn either left (a) or right (b).

increase uncertainty due to stiffness along the needle shaft. Medical imaging in the operating room can be used to measure the needle's current position and orientation to provide feedback to the planner [55, 67], but this measurement by itself provides no information about the effect of future deflections during insertion due to motion uncertainty.

Stochastic motion roadmaps offer features particularly beneficial for medical needle steering. First, SMR's explicitly consider uncertainty in the motion of the needle. Second, intra-operative medical imaging can be combined with the fast SMR queries to permit control of the needle in the operating room without requiring time-consuming intra-operative re-planning.

6.2.1 SMR Implementation

We formulate the SMR for a bang-bang steering car, which can be applied to needle steering. The state of such a car is fully characterized by its position $\mathbf{p} = (x, y)$, orientation angle θ, and turning direction b, where b is either left ($b = 0$) or right ($b = 1$). Hence, the dimension of the state space is $d = 4$, and a state i is defined by $s_i = (x_i, y_i, \theta_i, b_i)$, as illustrated in figure 6.3. We encode b_i in the state since it is a type of history parameter that is required by the motion uncertainty model. Since an SMR assumes that each component of the state vector is a real number, we define the binary b_i as the floor of the fourth component in s_i, which we bound in the range $[0, 2)$.

Between sensor measurements of state, we assume the car moves a distance δ. The set U consists of two actions: move forward turning left ($u = 0$), or move forward turning right ($u = 1$). As the car moves forward, it traces an arc of length δ with radius of curvature r and direction based on u. We consider r and δ as random variables drawn from a given distribution. In this chapter, we consider $\delta \sim N(\delta_0, \sigma_{\delta_a})$ and $r \sim N(r_0, \sigma_{r_a})$, where N is a normal distribution with given mean and standard deviation parameters and $a \in \{0, 1\}$ indicates direction change. We implement `generateSampleTransition` to draw random samples from these distributions. Although the uncertainty parameters can be difficult to measure precisely, even rough estimates may be more realistic than using deterministic transitions when uncertainty is high.

We define the workspace as a rectangle of width x_{max} and height y_{max} and define obstacles as polygons in the plane. To detect obstacle collisions,

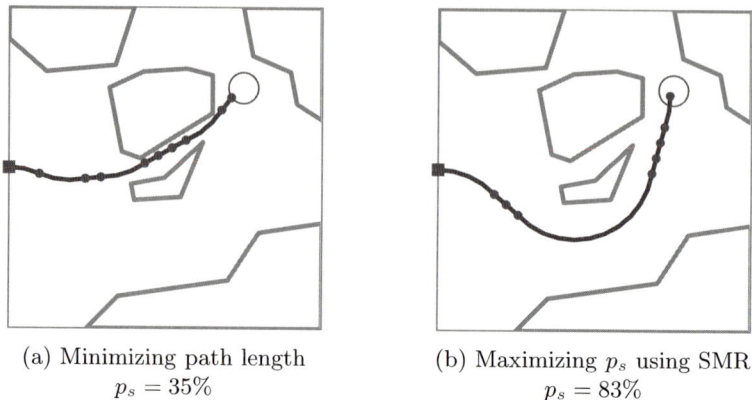

(a) Minimizing path length   (b) Maximizing p_s using SMR
$p_s = 35\%$ $p_s = 83\%$

Fig. 6.4. Explicitly considering motion uncertainty using an SMR planner improves the probability of success

we use the zero-winding rule [99]. We define the distance between two states s_1 and s_2 to be the weighted Euclidean distance between the poses plus an indicator variable to ensure the turning directions match: $\texttt{distance}(s_1, s_2) = \sqrt{(x_1 - x_2)^2 + (y_1 - y_2)^2 + \alpha(\theta_1 - \theta_2)^2} + M$, where $M \to \infty$ if $b_1 \neq b_2$, and $M = 0$ otherwise. For fast nearest-neighbor computation, we use the CGAL implementation of kd-trees [195]. Since the $\texttt{distance}$ function is non-Euclidean, we use the formulation developed by Atramentov and LaValle to build the kd-tree [23]. We define the goal T^* as all configuration states within a ball of radius t^r centered at a point $\mathbf{t}^*$.

6.2.2 Results

We implemented the SMR planner in C++ and tested the method on workspaces of size $x_{max} = y_{max} = 10$ with polygonal obstacles as shown in shown in figure 6.1 and figure 6.4. We set the robot parameters $r_0 = 2.5$ and $\delta_0 = 0.5$ with motion uncertainty parameters $\sigma_{\delta_0} = 0.1$, $\sigma_{\delta_1} = 0.2$, $\sigma_{r_0} = 0.5$, and $\sigma_{r_1} = 1.0$. We set parameters $\gamma = 0.00001$ and $\alpha = 2.0$. We tested the motion planner on a 2.2 GHz AMD Opteron PC. Building the SMR required approximately 1 minute for $n = 50,000$ states, executing a query required 6 seconds, and additional queries for the same goal required less than 1 second of computation time for both example problems.

We evaluate the plans generated by SMR with multiple randomized simulations. Given the current state of the robot, we query the SMR to obtain an optimal action u. We then execute this action and compute the expected next state. We repeat until the robot reaches the goal or hits an obstacle, and we illustrate the resulting expected path. Since the motion response of the robot to actions is not deterministic, success of the procedure can rarely be guaranteed. To estimate p_s, we run the simulation 100 times, sampling the next state from the transition probability distribution rather than selecting the expected

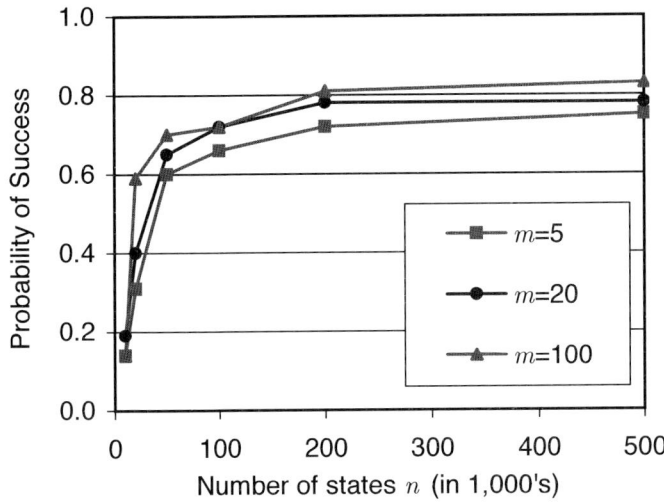

Fig. 6.5. Effect of the number of states n and number of motion samples m on the probability of success p_s

value, and we compute the number of goal acquisitions divided by the number of obstacle collisions.

In figure 6.4(b), we illustrate the expected path using an SMR with $m = 100$ motion samples and $n = 500,000$ states. As in figure 6.1(b), the robot avoids passing through a narrow passageway near the goal and instead takes a longer route. The selection of the longer path is not due to insufficient states in the SMR; there exist paths in the SMR that pass through the narrow gaps between the obstacles. The plan resulting in a longer path is selected purely because it maximizes p_s.

The probability of success p_s improves as the sampling density of the configuration space and the motion uncertainty distribution increase, as shown in figure 6.5. As n and m increase, $p_s(s)$ is more accurately approximated over the configuration space, resulting in better action decisions. However, p_s effectively converges for $n \geq 100,000$ and $m \geq 20$, suggesting the inherent difficulty of the motion planning problem. Furthermore, the expected path does not substantially vary from the path shown in figure 6.4(b) for $n \geq 50,000$ and $m \geq 5$. The number of states required by the SMR planner is far smaller than the 800,000 states required for a similar problem using a grid-based approach with bounded error [7].

In figure 6.4(a), we computed the optimal shortest path assuming deterministic motion of the robot using a fine regular discrete grid with 816,080 states for which the error due to discretization is small and bounded [7]. We estimate p_s using the same simulation methodology as for an SMR plan, except that we compute the shortest path for each query. The expected shortest path passes

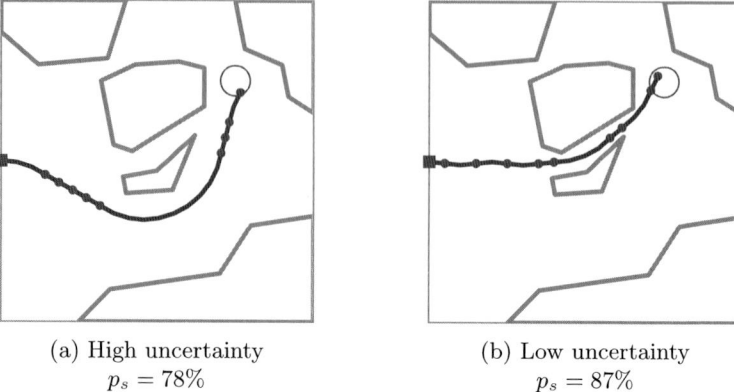

(a) High uncertainty
$p_s = 78\%$

(b) Low uncertainty
$p_s = 87\%$

Fig. 6.6. The level of uncertainty affects SMR planning results. In cases of low uncertainty (with 75% reduction in distribution standard deviations), the expected path resembles a deterministic shortest path due to the small influence of uncertainty on p_s and the effect of the penalty term γ. In both these examples, the same $n = 200,000$ states were used in the roadmap.

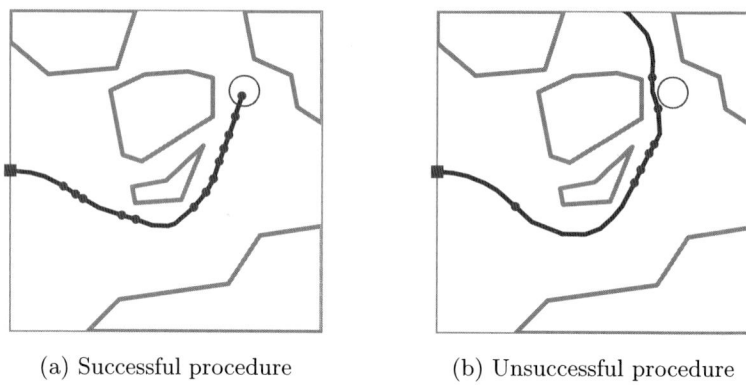

(a) Successful procedure

(b) Unsuccessful procedure

Fig. 6.7. Two simulated procedures of needle steering, one successful (a) and one unsuccessful due to effects of uncertain motion (b), using an SMR with $n = 50,000$ states

through a narrow passage between obstacles and the resulting probability of success is substantially lower compared to the SMR plan. The result was similar for the example in figure 6.1; explicitly considering motion uncertainty improved the probability of success.

To further illustrate the importance of explicitly considering uncertainty during motion planning, we vary the standard deviation parameters σ_{δ_0}, σ_{δ_1}, σ_{r_0}, and σ_{r_1}. In figure 6.6, we compute a plan for a robot with each standard deviation parameter set to a quarter of its default value. For this low uncertainty case, the uncertainty is not sufficient to justify avoiding the narrow passageway; the penalty γ causes the plan to resemble the deterministic shortest plan in figure 6.4(a). Also, p_s is substantially higher because of the lower uncertainty.

In figure 6.7, we execute the planner in the context of an image-guided procedure. We assume the needle tip position and orientation is extracted from a medical image and then execute a query, simulate the needle motion by drawing a sample from the motion uncertainty distribution, and repeat. The effect of uncertainty can be seen as deflections in the path. In practice, clinicians could monitor $p_s(s)$ for the current state s as the procedure progresses.

6.3   Conclusion and Open Problems

In many motion planning applications, the response of the robot to commanded actions cannot be precisely predicted. We introduce the Stochastic Motion Roadmap (SMR), a new motion planning framework that explicitly considers uncertainty in robot motion to maximize the probability that a robot will avoid obstacle collisions and successfully reach a goal. SMR planners combine the roadmap representation of configuration space used in PRM with the theory of MDP's to explicitly consider motion uncertainty at the planning stage.

To demonstrate SMR's, we considered a nonholonomic mobile robot with bang-bang control, a type of Dubins-car robot model that can be applied to steering medical needles through soft tissue. Needle steering, like many other medical procedures, is subject to substantial motion uncertainty and is therefore ill-suited to shortest-path plans that may guide medical tools through narrow passageways between critical tissues. Using randomized simulation, we demonstrated that SMR's generate motion plans with significantly higher probabilities of success compared to traditional shortest-path approaches.

Because SMR is a framework, extensions to the underlying methods can be applied to wide variety of problems. Several extensions that would expand the applicability of SMR include considering actions in a continuous range rather than solely from a discrete set, investigating more sophisticated sampling methods for generating configuration samples and for estimating transition probabilities, and integrating the effects of sensing uncertainty. We hope that SMR can be applied to new biomedical and industrial problems where uncertainty in motion should be considered to maximize the probability of success.

7 Motion Planning for Radiation Sources during High-Dose-Rate Brachytherapy

High-dose-rate (HDR) brachytherapy is a type of radiation treatment for cancer. In this procedure, a physician guides radioactive sources through catheters that have been inserted inside or near the cancerous tumors. The goal is to provide a high radioactive dose to treat the tumor while not significantly damaging surrounding healthy tissues. HDR brachytherapy has been successfully used for treating many types of cancer, including prostate cancer [176], cervical cancer [145], and breast cancer [107].

When treating cancer using radiation, physicians desire dose distributions that conform to patient anatomy and satisfy dose prescriptions for the tumor target and nearby critical organs [106]. Using medical images of patient anatomy and estimates of tumor location, physicians prescribe radiation dose requirements for cancerous tumors and surrounding tissues. A sample slice of a CT scan used for this purpose for a prostate cancer patient case is shown in figure 7.1. The goal is then to move the radioactive source inside the catheters to generate a dose distribution that satisfies the clinical criteria as best as possible. This goal can be formulated as an optimization-based motion planning problem: how should we move the radioactive seed through the catheters such that the dose delivered to the patient minimizes the deviation from the prescribed dose?

We draw on linear programming to develop a fast and exact method to optimize radioactive source locations and dwell times for HDR brachytherapy cancer treatment. The method uses the objective and clinical criteria framework of Inverse Planning by Simulated Annealing (IPSA), an approach developed by Lessard and Pouliot in 2000 that has been used in the treatment of over a thousand patients [144, 145, 146]. By formulating the HDR brachytherapy dose optimization problem as a linear program, we enable the fast computation of mathematically optimal solutions.

In this chapter, we present our linear programming formulation and apply the method to a sample of 20 prostate cancer patient cases [13, 14]. We then quantitatively compare the mathematically optimal dwell times solutions for HDR brachytherapy treatment obtained using the LP method to the solutions currently being obtained clinically using simulated annealing (SA), a probabilistic

R. Alterovitz and K. Goldberg: Motion Planning in Medicine, STAR 50, pp. 91–106, 2008.
springerlink.com   © Springer-Verlag Berlin Heidelberg 2008

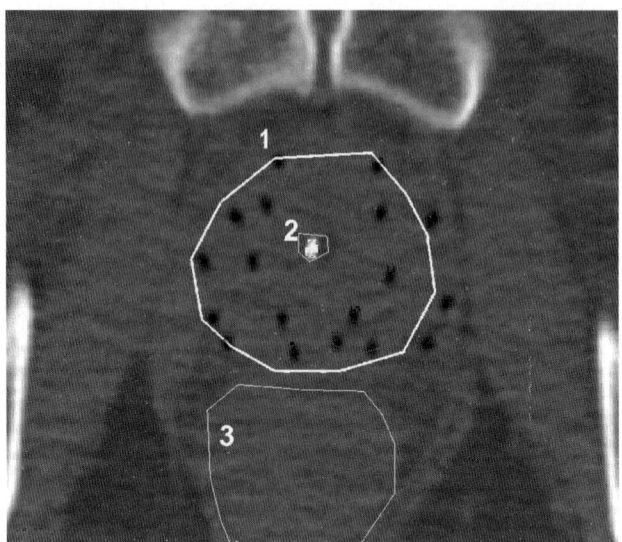

Fig. 7.1. Transverse slice of a CT scan with white contours of the prostate (1), urethra (2), and rectum (3). The catheters are marked with black dots.

method that is not guaranteed to return an optimal solution in finite computation time. We show that the LP method resulted in significantly improved objective function values compared to SA, but the dose distributions produced by the dwell times solutions were clinically equivalent as measured by standard dosimetric indices with a 2% threshold.

7.1   Introduction to HDR Brachytherapy and Dose Optimization

The specifics of the the HDR brachytherapy procedure depend on the cancer site. For the case of the HDR brachytherapy for prostate cancer, the physician commonly implants 14 to 18 catheters in the prostate through the perineum under ultrasound guidance. The physician obtains an image (usually CT scan or MRI) of the catheters and the surrounding tissue, which is used to specify dose prescriptions for the patient anatomy. The catheters are then attached to an HDR Remote Afterloader for treatment delivery. The afterloader, a type of robot, moves a single radioactive source, typically 4.5 mm long and 0.9 mm in diameter containing ^{192}Ir, inside each catheter.

In clinical practice, the seed is generally not moved at a continuous speed through the catheters but rather is temporarily stopped at predetermined dwell positions. Between stops, the seed moves at high speed. The use of predetermined dwell locations converts the problem from a motion planning problem in continuous space to a discrete optimization problem. By adjusting the length of time (dwell time) that the source remains at any location within a catheter (dwell position), it is possible to generate a wide variety of dose distributions.

To address the dose optimization problem, Lessard and Pouliot developed Inverse Planning by Simulated Annealing (IPSA) [144, 145, 146]. IPSA has been used in the treatment planning of over a thousand patients at UCSF since 2000 and has been independently evaluated by several American and European institutions [54, 64, 132, 150, 154, 197].

A complete description of IPSA and its clinical applications was recently published [177]. Only the elements required for the present work are described here. Using hand-segmented boundaries of the dominant intraprostatic lesions and nearby organs [176], the software generates a discrete sample of dose calculation points inside and on the boundary of the tissue types. For dose calculation points of each tissue type, IPSA permits the physician to prescribe unique dose ranges as well as penalty costs that grow linearly when actual dose violates the prescribed dose ranges. Setting dwell times to minimize dose penalty costs rather than using rigid dose constraints guarantees that the method will find an achievable solution. IPSA defines an objective function equal to a weighted sum of penalty costs at dose calculation points given the dwell times. In the IPSA framework, the mathematically optimal solution is the solution of dwell times that globally minimizes the objective function. IPSA's single objective function assumes that the clinician has specified desirable dose penalty costs and generates a single dwell times solution, in contrast to multi-objective optimization formulations that consider the weights as variables and generate a Pareto front of solutions [134, 135].

The current version of IPSA software uses simulated annealing (SA) to compute dwell times to minimize the objective function. The computation time for a typical case is about 10 seconds on PC with a 3.6 GHz Intel Xeon processor (Nucletron's Masterplan Station). The computation time includes the automatic selection of the active dwell positions, the generation of the dose calculation points, the generation of a look-up dose-rate table, and 100,000 simulated annealing iterations. SA applies a random search with the ability to escape local minima and offers a statistical guarantee to converge asymptotically to the global minimum [1, 89, 205]. The longer the SA algorithm searches for a solution, the higher the probability that the optimal solution is found. Although this method has worked well in clinical practice using 100,000 iterations, there previously was no general quantitative information available regarding the closeness to mathematical optimality of the solutions obtained using simulated annealing, a probabilistic method that cannot guarantee the achievement of a global minimum within a finite computation time.

7.2 Linear Programming for HDR Brachytherapy

Our primary contribution is to take the well-established dose optimization problem defined by IPSA and show that it can be exactly formulated as a linear programming (LP) problem. Because the global minimum for an LP problem can be computed exactly and deterministically using pre-existing algorithms, this formulation provides strong performance guarantees for cost minimization: one can rapidly find the minimum cost solution for any patient case and clinical criteria parameters. LP does not require setting parameters specific to the optimization method, such as stopping criteria or pseudo-temperatures for SA or mutation probabilities for GA [134, 135, 221]. This allows clinicians to customize dose prescriptions and penalty costs based on medical considerations without concern about their effect on the convergence of the optimization method. Unlike other deterministic algorithms such as local search [146], the LP method will never be trapped at sub-optimal solutions of IPSA's objective function. Since the LP solution is guaranteed to globally minimize the objective function, it provides a precise baseline for evaluating solutions currently being obtained clinically by probabilistic methods such as SA.

Our second contribution is to quantitatively compare the dwell times solutions for HDR treatment currently being obtained clinically using simulated annealing (SA) to the mathematically optimal solutions obtained using LP. With a sample of 20 prostate cancer patient cases, we show that the LP method resulted in significantly improved objective function values compared to SA, but the dose distributions produced by the dwell times solutions were clinically equivalent as measured by standard dosimetric indices.

A linear programming problem is defined by an objective function and constraints that are linear functions of the variables. An LP problem can be solved using the Simplex algorithm, a global deterministic optimization method that considers the geometric polyhedron defined by the linear constraints and systematically moves along edges of the polyhedron to new feasible solutions (represented as vertices of the polyhedron) with successively better values of the objective function until the optimum is reached [161]. In 1990, Renner et al. was the first group to propose a linear programming formulation for HDR brachytherapy dose optimization. Their method minimizes the time the source is irradiating tissue subject to a minimum dose constraint for a set of points in the target volume [179]. Kneschaurek et al. extended this method to permit the specification of dose ranges using rigid constraints for both minimum and maximum dose[123]. Jozsef et al. also used rigid constraints on dose range and minimized the maximum deviation from a prescribed dose constant at dose calculation points [114]. However, a solution of dwell times that results in a dose distribution that satisfies the rigid constraints may not be physically realizable. By defining the dwell times as variables and defining rigid linear constraints on dose, these previous approaches formulated the LP problem in a manner that does not guarantee the output of a solution since no feasible solution may exist. Finding a clinically realizable solution in such cases necessitates arbitrarily removing some rigid dose constraints, which requires substantial human intervention.

Our new linear programming (LP) formulation combines the advantages of IPSA's cost functions and extensive clinical validation with the benefits of deterministic global optimization for cost minimization. We show that the new LP method computes in finite time the mathematically optimal solution for dwell times to generate the best achievable dose distribution given the clinical objectives and the pre-optimization data generated by IPSA (active dwell positions, dose calculation points, and dose rate look-up table). We applied both SA and the new LP method to 20 prostate cancer patient cases and evaluated improvement of results using objective function values and standard dosimetric indices.

7.2.1   Patient Data Input

The input to the method are 3-D images of the tissues surrounding the tumor. We assume anatomical structures corresponding to b tissue types are segmented, including the clinical target volume (CTV) and critical organs (CO). We also assume the catheters are segmented. From the segmented anatomical structures, we use IPSA to select the active dwell positions and generated a set of m dose calculation points for which the optimization methods will calculate dose. The dose calculation points are distributed based on the anatomy and the implant in order to represent an accurate measurement of the clinical objectives [144]. For each contoured volume, IPSA uses two categories of dose calculation points: "surface" and "volume." This results in $q = 2b$ dose calculation point types: "surface" and "volume" for the b segmented tissue types. For each tissue type, adjusting the dose to "surface" dose calculation points controls the dose coverage and conformality while adjusting the dose to "volume" dose calculation points controls the dose homogeneity [106].

7.2.2   Dose Calculation

Dwell positions are defined as points along catheters at which a source can be placed for a non-zero interval of time. The n active dwell positions were selected by IPSA. We define the dwell time of a source at dwell position j by t_j. A dwell time of 0 corresponds to skipping past a dwell position. The dwell times t_j are the variables that will be set to produce a dose distribution that satisfies the clinical criteria as best as possible.

We calculate the dose-rate contribution d_{ij} of a dwell position j to a dose calculation point i as specified in the AAPM TG-43 dosimetry protocol [162, 180]. The dose-rate contribution is the energy imparted by the radioactive source into an absorbing material (the tissue) per unit time and has units cGy/sec, where 1 gray (Gy) equals 1 joule per kilogram. The dose-rate contribution is a function of r_{ij}, the distance between the dwell position j and the dose calculation point i. It also depends on the radioactive material used in the source, which in this case was ^{192}Ir. Since small differences in the dose calculation may affect the outcome of the optimization, we use the look-up dose-rate table calculated by IPSA as an input for the LP method.

The dose contribution of a dwell position j to a dose calculation point i is computed by multiplying the dose-rate contribution d_{ij} by the dwell time t_j.

The dose D_i at a dose calculation point i, which has units of cGy, is calculated by summing the dose contribution from each dwell position.

$$D_i = \sum_{j=1}^{n} d_{ij} t_j.$$

The dose D_i has units of cGy, which describes the energy imparted by radiation into a unit mass of tissue.

7.2.3   Clinical Criteria

After contouring, the physician prescribes dose ranges for each anatomical structure. The dose ranges used in this study, listed in Table 7.1, are typical values clinically used at the UCSF Comprehensive Cancer Center for treating prostate cancer [106]. This includes the minimum dose D_s^{min} and maximum dose D_s^{max} for each dose calculation point type s. For a dose calculation point i of type s, the desired dose D_{si} should satisfy $D_s^{min} \leq D_{si} \leq D_s^{max}$.

Table 7.1. Clinical criteria parameters for dose penalty cost functions for a typical prostate cancer case

s	Dose calculation point type	D_s^{min} (cGy)	M_s^{min}	D_s^{max} (cGy)	M_s^{max}
1	Prostate (surface)	950	100	1425	100
2	Prostate (volume)	950	100	1425	30
3	Urethra (surface)	950	100	1140	30
4	Urethra (volume)	950	100	1140	30
5	Rectum (surface)	0	0	475	20
6	Rectum (volume)	0	0	475	20
7	Bladder (surface)	0	0	475	20
8	Bladder (volume)	0	0	475	20

In practice, it may not be physically possible to provide a radioactive dose in the physician specified range for every dose calculation point in the 3-D volume. Hence, the physician also specifies a "penalty" for any point for which the clinical criteria is not satisfied. If the actual dose is below or above the prescribed range, the penalty increases linearly at rates M_s^{min} and M_s^{max}, respectively. Adjustment of M_s^{min} and M_s^{max} sets the relative importance of dose range satisfaction between anatomical structures. The penalty weights M_s^{min} and M_s^{max} listed in Table 7.1 are typical values used at the UCSF Comprehensive Cancer Center for prostate cancer cases [106]. The penalty w_{si} at a dose calculation point i of type s can be described in mathematical form using a cost function.

$$w_{si} = \begin{cases} -M_s^{min}(D_{si} - D_s^{min}) & \text{if } D_{si} \leq D_s^{min} \\ M_s^{max}(D_{si} - D_s^{max}) & \text{if } D_{si} \geq D_s^{max} \\ 0 & \text{if } D_s^{min} < D_{si} < D_s^{max} \end{cases} \tag{7.1}$$

Figure 7.2 plots the cost functions (penalty as a function of dose) for the prostate cancer clinical criteria in Table 7.1.

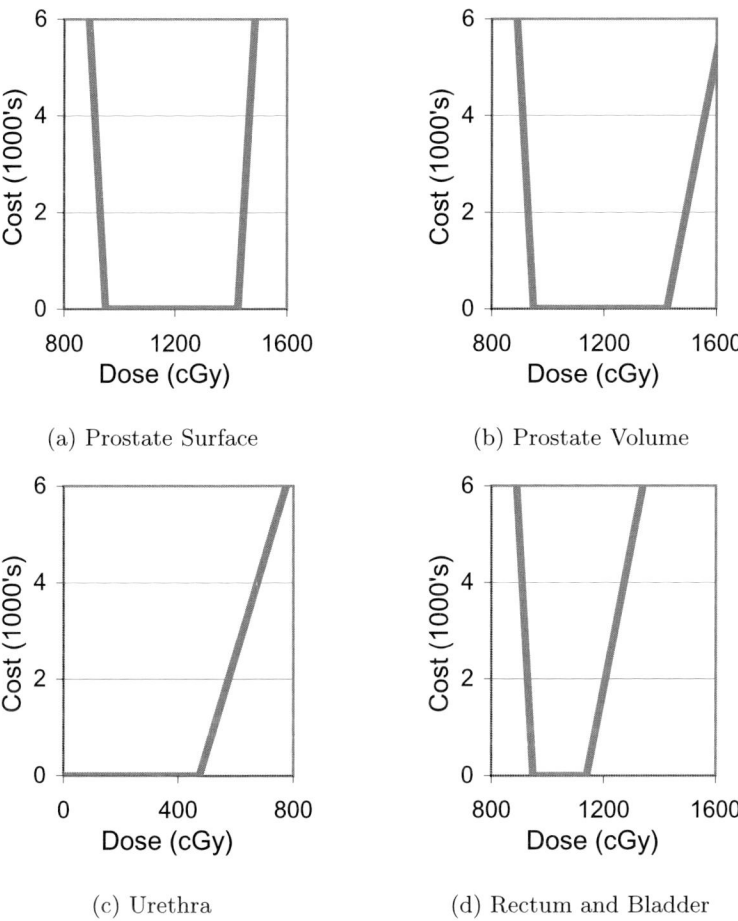

(a) Prostate Surface

(b) Prostate Volume

(c) Urethra

(d) Rectum and Bladder

Fig. 7.2. The clinical criteria, plotted here for a typical prostate cancer case, are specified using cost functions which define penalty as a function of dose for each dose calculation point type

7.2.4 Linear Programming Formulation

The objective is to satisfy the clinical criteria as best as possible by computing dwell times that minimize the net dose penalty costs. Equation (7.1) from section 7.2.3 defines the cost function for an individual dose calculation point i of type s based on the clinical criteria for that point. For each type s, we define the penalty cost E_s as the average penalty cost per point:

$$E_s = \sum_{i=1}^{m_s} \frac{w_{si}}{m_s} \tag{7.2}$$

where m_s is the number of dose calculation points of type s. The objective function E is effectively a weighted sum of the average cost for each tissue type s, where the relative weights are determined by the costs M_s^{min} and M_s^{max}. The global objective function is to minimize the sum of the penalty costs for the q dose calculation point types:

$$E = \sum_{s=1}^{q} E_s = \sum_{s=1}^{q} \sum_{i=1}^{m_s} \frac{w_{si}}{m_s}. \qquad (7.3)$$

This objective function is identical to the objective function used by IPSA [146].

The objective function E is not linear because it is composed of nonlinear functions w_{si}. However, each function w_{si} is piece-wise linear. We can formulate this problem as a linear program by creating artificial variables c_{si} to represent cost and defining the following constraints:

$$\begin{aligned} c_{si} &\geq -M_s^{min}(D_{si} - D_s^{min}) \\ c_{si} &\geq M_s^{max}(D_{si} - D_s^{max}) \\ c_{si} &\geq 0. \end{aligned} \qquad (7.4)$$

Because w_{si} is a piece-wise linear and convex function, the constraints above guarantee that $c_{si} \geq w_{si}$ for all i, s. Furthermore, we redefine the global objective function to

$$E = \sum_{s=1}^{q} \sum_{i=1}^{m_s} \frac{c_{si}}{m_s}. \qquad (7.5)$$

For minimized E where the costs c_{si} satisfy the inequalities (7.4), we are guaranteed $c_{si} = w_{si}$ for all s, i. We show this by proving the contrapositive ($c_{si} \neq w_{si}$ implies E not minimized), which is logically equivalent [198]. If $c_{si} \neq w_{si}$, then $c_{si} > w_{si}$ for some s, i and there will exist a cost c'_{si} such that $c_{si} > c'_{si} \geq w_{si}$. Since c'_{si} will not violate any constraint in inequalities (7.4), it is feasible. We define E' exactly as E except using c'_{si} instead of c_{si}. Hence, $E' < E$ and no cost variables used to compute E' violate a constraint, which implies E is not minimized. Hence, for minimized E, we are guaranteed $c_{si} = w_{si}$.

Table 7.2. HDR dose optimization LP formulation constants, variables, and functions

Constants:	
m_s	Number of dose calculation points of type s.
N	Number of dwell positions.
d_{sij}	Dose-rate contribution from dwell position j to dose calculation point i of type s.
Variables:	
t_j	Source dwell time for dwell position j.
c_{si}	Penalty cost at dose calculation point i of type s.
Objective:	
E	Global cost function.

We summarize constants, variables, and the objective function for the LP formulation in Table 7.2. In equation (7.6), we explicitly define the linear program in canonical form [161] by plugging into the constraints the dose distribution D_{si} at point i of type s due to dwell times t_j.

$$\text{Minimize} \quad E = \sum_{s=1}^{q} \sum_{i=1}^{m_s} \frac{c_i}{m_s}$$

Subject to:

$$c_{si} + \sum_{j=1}^{n} M_s^{min} d_{sij} t_j \geq M_s^{min} D_s^{min} \quad s = 1, \ldots, q; i = 1, \ldots, m_s$$

$$c_{si} - \sum_{j=1}^{n} M_s^{max} d_{sij} t_j \geq -M_s^{max} D_s^{max} \quad s = 1, \ldots, q; i = 1, \ldots, m_s$$

$$c_{si} \geq 0 \quad s = 1, \ldots, q; i = 1, \ldots, m_s$$

$$t_j \geq 0 \quad j = 1, \ldots, n$$

$$(7.6)$$

A (non-optimal) feasible solution for the LP formulation can be trivially found by setting $t_j = 0$ for all j and setting

$$c_{si} = \max\{ \; -M_s^{min} \left(\left(\sum_{j=1}^{n} d_{sij} t_j \right) - D_s^{min} \right),$$

$$0,$$

$$M_s^{max} \left(\left(\sum_{j=1}^{n} d_{sij} t_j \right) - D_s^{max} \right) \}$$

for all i and s.

Because of the properties of the artificial variables c_{si} shown above for minimized E, the optimal solution obtained for the linear program in equation (7.6) will be the same as the optimal solution to the nonlinear formulation based on the objective function in equation (7.3) with the cost functions in equation (7.1). We effectively transformed the nonlinear IPSA optimization problem in equation (7.3) (for which deterministic optimization algorithms such as local search could be trapped in sub-optimal solutions [146]) to a higher dimensional space with artificial variables in which an equivalent linear formulation (7.6) can be minimized deterministically to find the global optimal solution using the Simplex algorithm.

7.3 Application to Prostate Cancer Treatment

We implemented software using C++ to read patient specific parameters from IPSA and output the linear program (7.6) in the file format of AMPL (A Mathematical Programming Language) [85]. We solved the linear program specified in each AMPL file using ILOG CPLEX 9.0, an advanced implementation of the Simplex algorithm [161] designed for large industrial optimization problems [109]. Computation was performed on a 3.0 GHz Pentium IV computer running the Linux operating system.

7.3.1   Patient Data Sets

We applied the LP method retrospectively to 20 prostate cancer patient cases. The prostate volumes ranged from 23 cc to 103 cc. For these patients, the physician implanted 14 to 18 catheters in the prostate with transrectal ultrasound (TRUS) guidance while the patient was under epidural anesthesia. Then Flexiguide catheters (Best Industries, Inc., Flexi-needles, 283-25 (FL153-15NG)), which are 1.98 mm diameter hollow plastic needles through which the radioactive source moves, were inserted transperineally by following the tip of the catheter from the apex of the prostate to the base of the prostate using ultrasound and a stepper. A Foley catheter was inserted to help visualize the urethra.

After catheter implantation, a treatment planning pelvic CT scan was obtained for each patient. Three-millimeter-thick CT slices were collected using a spiral CT. The clinical target volume (CTV) and critical organs (CO) including bladder, rectum, and urethra were contoured using the Nucletron Plato Version 14.2.6 (Nucletron B.V., Veenendaal, The Netherlands). The CTV included only the prostate and no margin was added. When segmenting the bladder and rectum, the outermost mucosa surface was contoured. The urethra was defined by the outer surface of the Foley catheter. Only the urethral volume within the CTV was contoured. The CO's were contoured on all CT slices containing the CTV and at least two additional slices above and below. Implanted catheters were also segmented. A slice of a 3-D CT scan with contoured anatomical structures (prostate, urethra, rectum, and bladder) and catheters is shown in figure 7.1.

For the 20 cases, the number of dose calculation points m ranged from 1781 to 3510. Since the selection of the active dwell positions and dose calculation points affects the outcome of optimization [133], we use those generated by IPSA as input for the LP method. For the prostate cases, the images contained $q = 8$ dose calculation point types: "surface" and "volume" for the four contoured tissue types (prostate, urethra, bladder, and rectum). The clinical criteria used in this study are shown in 7.1.

All patients were treated at UCSF Comprehensive Cancer Center using dosimetric plans generated by the current version of IPSA. We used imaging and dosimetry records from those treatments to compare SA with LP.

7.3.2   Evaluation Metrics

We recorded the dwell times and the objective function value E for the solutions obtained using SA and LP. We evaluated the resulting dose distributions using standard dosimetric indices, including prostate V100 and V150 (the percentage of the prostate receiving over 100% and over 150% of the prescribed dose, respectively). As dose inside the prostate should fall between 100% (D^{min}) and 150% (D^{max}) of prescribed dose, ideally V100 should be 100% and V150 should be 0%. Similarly, we also evaluated V100 and V150 for the urethra. Dosimetric indices for normal structures (non-cancerous tissues) include the rectum (V50 and V100) and the bladder (V50 and V100). Because normal structures should be spared radioactive dose, these indices ideally should be close to 0%. We also computed dosimetric indices in absolute dose, including the prostate D90 (the maximal dose that covers 90% of prostate volume), urethra D10 (the maximal dose that covers 10% of urethra volume), and rectum and bladder D2cc (the maximal dose that covers 2 cc of the organ volume).

7.3.3   Results

ILOG CPLEX solved for the optimal solution to the linear programming formulation in an average time of 9.00 seconds per case with a standard deviation of 3.77 seconds for the 20 prostate cancer patient cases. The times ranged from 3.68 seconds to 14.63 seconds. The Simplex algorithm in ILOG CPLEX required an average of 1653 iterations with a standard deviation of 341 iterations.

The average objective function value for the 20 prostate cancer patient cases was 3.27 for the LP method compared to 3.33 for SA. The percent difference in objective function value between the solution found using SA and the optimal solution found using LP for each individual patient case is shown in figure 7.3.

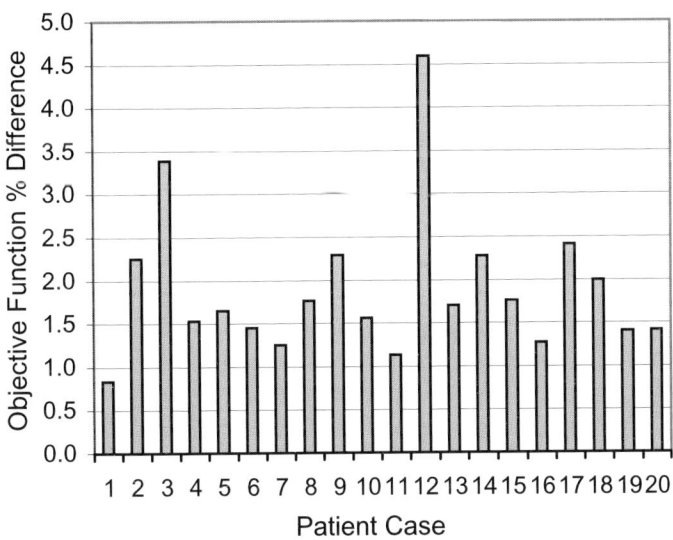

Fig. 7.3. The percent difference in objective function value between the optimal solution (found using the LP method) and the solution found by SA for 20 prostate cancer patient cases. The difference is statistically significant ($P = 1.54 \times 10^{-7}$).

Table 7.3. Improvement of LP solutions over SA solutions for 20 prostate cancer patient cases calculated as the absolute difference in dosimetric index percent values. Negative values indicate deterioration in the dosimetric index. The significance P of the differences was computed using paired t-tests.

Dosimetric Index	Maximum Improvement	Minimum Improvement	Mean Improvement	99% CI	Significance P
Prostate V100	0.95	-0.49	0.13 (-0.10, 0.37)		0.1644
Prostate V150	1.65	-1.63	0.51 (-0.02, 1.04)		0.0217
Urethra V100	1.52	-1.50	0.12 (-0.33, 0.57)		0.4858
Urethra V150	0.11	-0.05	0.00 (-0.01, 0.02)		0.7621
Rectum V50	0.50	-0.81	-0.17 (-0.36, 0.02)		0.0344
Rectum V100	0.03	0.00	0.01 (-0.00, 0.01)		0.0289
Bladder V50	0.75	-0.48	0.03 (-0.17, 0.23)		0.7042
Bladder V100	0.13	-0.02	0.02 (-0.00, 0.04)		0.0225

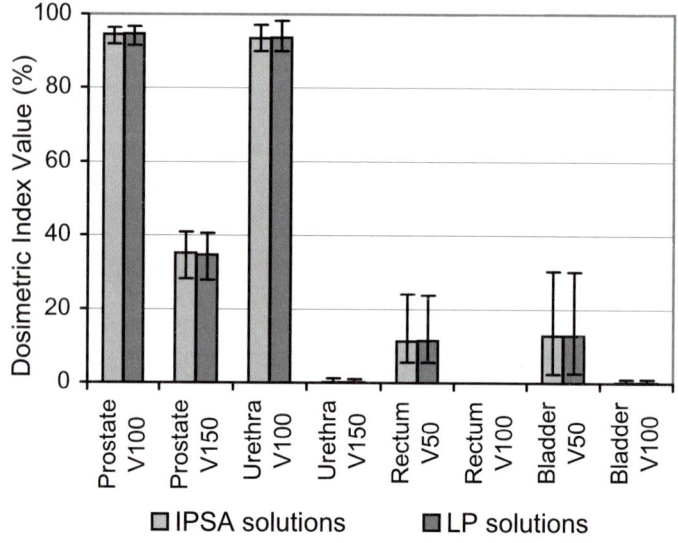

Fig. 7.4. Mean dosimetric index results for the SA and LP methods for 20 prostate cancer patient cases. Error bars indicate maximum and minimum values for the 20 patient cases.

Improvement varies from a minimum of 0.84% to a maximum of 4.59%. We performed paired t-tests to determine the statistical significance ($P < 0.01$) of the results and found that the improvement in objective function value using the LP method compared to SA was statistically significant ($P = 1.54 \times 10^{-7}$).

Figure 7.4 displays the standard dosimetric indices for both the SA and LP solutions. The bars indicate the mean indices as percents and the error bars indicate the maximum and minimum indices obtained for the 20 prostate cancer

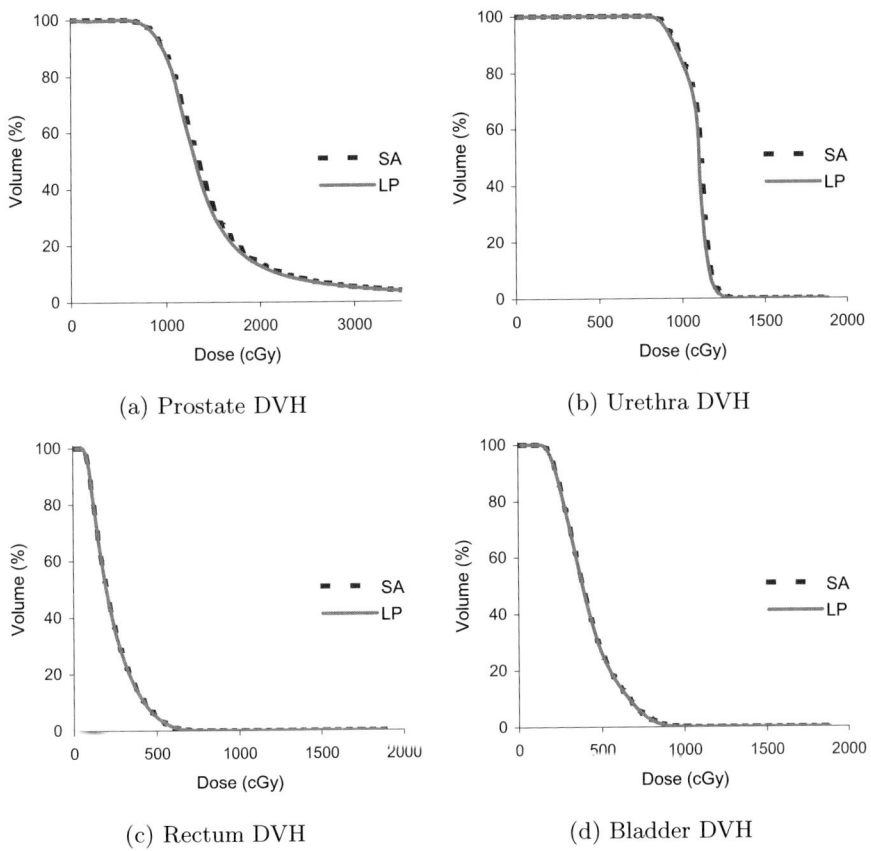

(a) Prostate DVH

(b) Urethra DVH

(c) Rectum DVH

(d) Bladder DVH

Fig. 7.5. Dose-volume-histogram (DVH) plots for the prostate (a), urethra (b), rectum (c), and bladder (d) for the patient case with greatest difference in dosimetric indices between the LP and SA solutions. For dose less than D^{min} for each tissue type, the desired volume is 100%. For dose greater than D^{max}, the desired volume 0%.

patient cases. Based on these dosimetric indices, the difference between the dose distributions generated by SA and LP was small. None of the dosimetric indices indicated a statistically significant ($P < 0.01$) difference between the dose distributions generated by SA and LP. The largest improvement for the prostate D90, the rectum D2cc, and the bladder D2cc were lower than 1%. The largest improvement for the urethra D10 was 2%. The urethra V150 was zero for both LP and SA method for this case. Additional dosimetric indices are shown in Table 7.3 where positive values indicate improvement and negative values indicate deterioration. The deterioration of one dosimetric index is sometimes traded for the improvement of other dosimetric indices and the improvement of the global solution. The maximum improvement of LP over SA was a reduction of 1.65% for the prostate V150 index. However, for the same patient, LP resulted in a reduction of 0.38% of the prostate V100. Similarly, the maximum deterioration

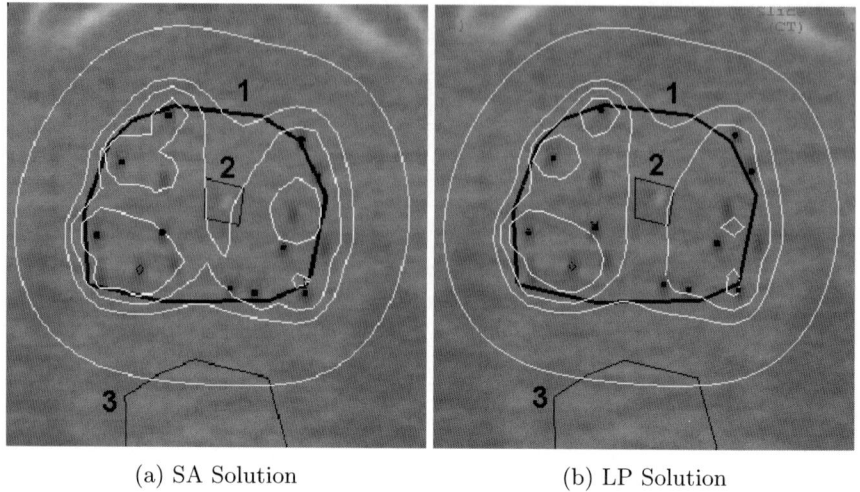

(a) SA Solution   (b) LP Solution

Fig. 7.6. Isodose curves for the SA (a) and LP (b) solutions for the patient case with greatest difference in dosimetric indices. The prostate (1), urethra (2), and rectum (3) are contoured in black. Catheters are shown as black dots. Isodose curves for 50%, 100% (D^{min}), 120%, and 150% (D^{max}) of prostate minimum prescribed dose are plotted in white.

of LP over SA was an increase of 1.63% for the prostate V150 index inducing an improvement of 0.84% of the prostate V100. Even with these two extreme cases, the LP and SA methods provide two different solutions that are difficult to distinguish clinically. Figure 7.5 plots the dose-volume-histogram (DVH) for each tissue type for the patient case with the greatest magnitude improvement in a dosimetric index between the SA and LP solutions. Figure 7.6 displays a CT scan of the same patient with overlaid isodose contours for both solutions.

7.4 Discussion

The dosimetric index results are not significantly different from those of the current version of IPSA, which was previously shown to be superior to the commonly used method of geometric optimization followed by manual adjustment [106, 132]. The small variances observed for the prostate and urethra in figure 7.4 show the consistency of the treatment plan quality for both the SA and LP methods. The larger variances for the prostate V150, the rectum, and the bladder are due to differences between patients in anatomy, prostate volume, and distances between the prostate and organs at risk.

The LP and SA methods are both based on IPSA's objective function for the HDR brachytherapy dwell time optimization problem. The only difference is the optimization algorithm used, simulated annealing versus an equivalent linear programming formulation that can be solved using the Simplex algorithm. As simulated annealing is a probabilistic method, it is only guaranteed to converge to an optimal solution after an infinite amount of computation time. Standard termination criteria, such as stopping the algorithm after a fixed number of iterations, can result in sub-optimal solutions. During the development phase of the current version of IPSA, a large number of cases were run using a very large number of iterations (>1 million) and no significant improvements in the dosimetric indices were found compared to the values found after 100,000 iterations. However, the closeness to mathematical optimality of the solutions of the current version of IPSA could not be guaranteed for every new clinical case.

Because the LP formulation of IPSA's objective function can be solved deterministically to find the solution that globally minimizes costs, the LP method solution provides a precise baseline for evaluating solutions obtained by probabilistic methods such as SA. The LP method computed a solution with a better objective function value compared to SA for every patient case. The improvement in objective function values of LP compared to SA was statistically significant. However, the effect size of the objective function improvement was not sufficient to result in statistically significant differences in standard dosimetric indices for our sample of 20 prostates with volume ranging from 23 cc to 103 cc. We observe that the DVH plots for the patient case with the largest difference in dosimetric indices are similar for both methods (figure 7.5) while differences are observable on the isodose curves (figure 7.6). The hot spots (prostate V150) have different shapes and the prostate V120 curve is at a different location. This indicates that the local dose distribution (isodose) is different while global dose delivered to the organs (DVH) and critical dose delivered to the organs (dosimetric indices) are equivalent. This quantitatively indicates that the dose distributions generated by SA are clinically equivalent to the best achievable dose distributions based on the current IPSA objective function with dose constraints and penalty weights selected for prostate cancer cases.

7.5 Conclusion and Open Problems

HDR brachytherapy requires that clinicians solve a motion planning problem: how should a radioactive source move through pre-implanted catheters to deliver the best achievable dose to the patient? This problem can be formulated as an optimization-based motion planning problem: set dwell times for the radioactive source at dwell positions along the catheters such that the resulting dose distribution minimizes the deviation from physician-specified dose prescriptions. The primary contribution of this chapter is to take the well-established dwell times optimization problem defined by Inverse Planning by Simulated Annealing (IPSA) developed at UCSF and exactly formulate it as a linear programming (LP) problem. Because LP problems can be solved exactly and deterministically, this formulation provides strong performance guarantees: one can rapidly find the dwell times solution that globally minimizes IPSA's objective function for any patient case and clinical criteria parameters. For a sample of 20 prostate cancer patient cases, the new LP method optimized dwell times in less than 15 seconds per case on a standard PC.

We quantitatively compared the dwell times solutions currently being obtained clinically using simulated annealing (SA), a probabilistic method, to the mathematically optimal solutions obtained using the LP method. The LP method resulted in significantly improved objective function values compared to SA, but none of the dosimetric indices indicated a statistically significant difference. The results indicate that solutions generated by the current version of IPSA are clinically equivalent to the mathematically optimal solutions.

IPSA's objective function with dose constraints and penalty weights covers all organs and all clinical objectives so they can be optimized simultaneously. The physician can adjust the objectives for each optimization. However, if a particular set of objectives generates the desired results, then the same set of objectives can be used for optimization of clinically similar cases (i.e. prostate) without further adjustments. This set of objectives, commonly called a class solution, can be used as a starting point for every patient, significantly reducing the time needed to plan individual patient treatments.

Although we focused on the application of our linear programming formulation to prostate cancer patient cases, the mathematical formulation can also be applied to other cancer types for which HDR brachytherapy is used by incorporating different clinical parameters. The method can also be extended to support any piece-wise linear convex cost functions, not solely the 3-piece cost functions presented above. Recent developments in magnetic resonance spectroscopy imaging and image registration introduce a new clinical criterion, a dose boost to the tumor volume within the prostate [8, 11, 122]. Although we do not explicitly consider that dose calculation point type, the mathematical formulation we defined can be extended to incorporate it by adding a tumor volume tissue type. A potential advantage of the LP method for each of these extensions is that it will use the well-established framework of IPSA and deterministically compute mathematically optimal dwell time solutions for all patient cases.

Although we used linear programming purely as an optimization method in this chapter, linear programming brings with it a vast literature of tools and extensions. This includes well-established tools like sensitivity analysis [161], as well as newer extensions like robust optimization that considers uncertainty in input parameters [33]. In future work, we plan to explore these tools and extensions to provide clinicians with patient-specific information about the trade-offs between feasible treatment plans.

8 Conclusion

This monograph introduces a set of algorithms that computationally plan and optimize image-guided medical procedures based on imaging data and physician-specified clinical criteria. These computational methods bridge the gap between medical imaging, where emerging advancements are enabling clinicians to non-invasively examine anatomy and metabolic processes in detail, and medical robotics, which is rapidly gaining acceptance in clinical practice.

The monograph focuses on three motion planning problems that arise in image-guided medical procedures: motion planning for rigid needles, motion planning for steerable needles , and motion planning for radiation sources for cancer treatment. Each of these motion planning problems introduces new computational challenges and is subject to unique planning and optimization constraints imposed by the physician's treatment requirements, the patient's anatomy, and the physical limitations of medical equipment and devices. We present planning and optimization algorithms for each of these general problems, and then customize the solutions to the specific application of prostate brachytherapy.

8.1 Contributions

Motion planning for image-guided medical procedures presents three major algorithmic challenges: deformations, uncertainty, and optimality.

8.1.1 Deformations

When surgical devices such as needles contact soft tissue, the soft tissue may deform. Clinicians must compensate for these deformations to successfully guide a surgical device to a clinical target. To facilitate this, we propose optimization-based motion planning. The "cost" of a candidate plan is a function of the resulting placement error, obstacle collisions, and path length. This function can be evaluated by executing a physically-based simulation for a candidate plan. We can compute the optimal plan by using the physically-based simulation as a function in an optimization algorithm that minimizes cost.

R. Alterovitz and K. Goldberg: Motion Planning in Medicine, STAR 50, pp. 107–113, 2008.
springerlink.com

An essential component of this approach is a simulation of tissue deformations that occur when surgical devices such as needles contact soft tissue. We identified and implemented appropriate models and algorithms to interactively estimate soft tissue deformations due to forces applied during surgical and interventional medical procedures. The software tools integrate methods from real-time physically-based simulation in computer graphics and classical finite element methods.

8.1.2 Uncertainty

The motion response of surgical devices to commanded actions cannot be predicted with absolute certainty. Errors arise due to patient variability as well as limitations inherent to the surgical device (for example, a "rigid" needle flexing due to contact with tissue). Clinicians can take this uncertainty into account to guide surgical devices to a clinical target with a high probability of success.

This monograph presents the Stochastic Motion Roadmap (SMR), a new general motion planning framework that explicitly considers motion uncertainty during planning by combining motion sampling with Markov Decision Processes and Dynamic Programming. We applied the SMR framework to needle steering and showed that accounting for needle motion uncertainty during planning can significantly increase the probability of reaching targets without colliding with obstacles.

8.1.3 Optimality

Throughout this book, we focused on optimization-based motion planning. For needle insertion and needle steering, we minimized costs due to obstacle collisions, path length, and placement error. For needle steering with motion uncertainty, we maximized the probability of success. For radiation source motion planning, we minimized the deviation from physician-specified dose requirements. Optimization is a powerful framework for formulating and computing motion plans that maximize the probability of successfully achieving clinical goals while minimizing tissue damage and other negative side effects.

8.2 Future Directions

Advances in imaging and robotic surgical devices continue to introduce challenges and offer new opportunities for future research. In this section we outline some of these, including approaches for extending these results from 2-D planes to 3-D tissue volumes and new optimization-based planning approaches that can explicitly consider uncertainty in tool/tissue interaction.

8.2.1 Realistic Simulation of Image-Guided Medical Procedures

In chapters 2, 3, and 4, we focused on motion planning problems for image-guided medical procedures that consider tissue deformations on a 2-D imaging plane. Due to constraints of many imaging modalities such as ultrasound imaging, often only 2-D anatomical and tissue deformation information is available. However, the improving performance and increasing availability of full 3-D imaging modalities such as CT scans and MRI is introducing the ability to acquire 3-D patient-specific information pre- and intra-operatively. The 2-D simulations described in this book can serve as a foundation on which to develop accurate and efficient 3-D FEM simulations of tissue deformations.

The first challenge in simulating 3-D deformations is to define meshes of appropriate complexity to represent heterogeneous tissue volumes. For many image-guided procedures, the input for meshing will be a 3-D image volume and segmentation information. The segmentation information generally includes anatomical structures and regions of interest specified using polygonal outlines on multiple slices of the 3-D volume. From these outlines, it is possible to generate surface meshes for each tissue type using methods such as Marching Cubes [149]. Open source software tools such as NETGEN and TetGen [187, 193] can then be used to automatically generate a 3-D tetrahedral mesh from the tissue type surface meshes. Mesh decimation and smoothing may be required to reduce the number of elements; the goal is to generate a mesh that is sufficiently sparse to support fast FEM simulation while having sufficient density in key regions to realistically model the tissue. We illustrate 3-D surface meshes for the prostate and several surrounding tissues in figure 8.1.

The next challenge is representing forces exerted by the needle on the soft tissue. In 2-D, we modified the mesh as the needle was being inserted, maintaining mesh nodes along the needle shaft so that we could apply cutting and frictional forces as FEM boundary conditions. In 3-D, there are three approaches that should be considered: mesh modification, mesh refinement, and meshless methods. Although mesh modification worked well in our 2-D simulation in chapter 2, it is unclear whether modification of the 3-D mesh can be performed without resulting in degenerate elements (elements that invert and have "negative" area). With mesh refinement, new nodes are created in the vicinity of the needle path, which has been successfully applied to needle insertion in with regular meshes [165], but will require significant improvements in algorithmic computational complexity to be appropriate for real-time interactive simulation. Another potentially promising approach to explore is meshless methods, a relatively new approach based on clouds of linked nodes [148].

There is an inherent trade-off between the accuracy and speed of physically-based simulation algorithms for medical procedures involving soft tissue deformations. Methods like mass-spring systems introduced in chapter 2 achieve high speed (as measured by the frame rate of the simulation) while finite element methods used in chapters 3 and 4 achieve higher accuracy but are slower. This trade-off becomes more pronounced when transitioning from 2-D simulations to 3-D simulations due to the added computational complexity of computing deformations for 3-D models. As illustrated in figure 8.2, the appropriate method to select depends on the application; while simulation for physician training is subject to strict real-time performance requirements, accuracy is more critical for patient-specific procedure planning. The development of new physically-based

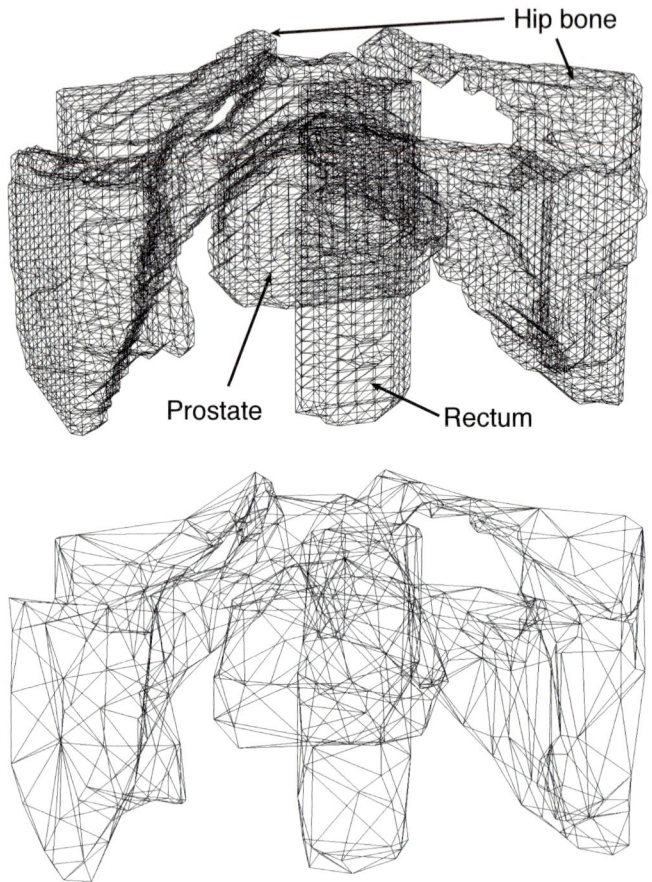

Fig. 8.1. We illustrate 3-D meshes generated from segmented slices from an MRI scan. The mesh on the top encodes the tissue geometry with high accuracy while the sparser mesh on the bottom is more suitable for real-time physically-based simulation.

simulation algorithms can push the trade-off frontier outwards, enabling faster, more accurate simulations of medical procedures.

For many types of soft tissue, the relationship between stress and strain is highly nonlinear [86]. Fast and accurate simulation of 3-D deformable soft tissues with nonlinear behavior has not yet been integrated into a practical surgery simulator. An interesting avenue for research is to explore computationally efficient methods to simulate realistic tissue mechanics and dynamics due to forces applied by surgical instruments or devices. In chapter 2, we focused on linearly elastic models of soft tissue. Future work should incorporate nonlinear tissue behavior, including both nonlinear geometry and nonlinear soft tissue material properties, to accurately simulate large deformations. To more realistically

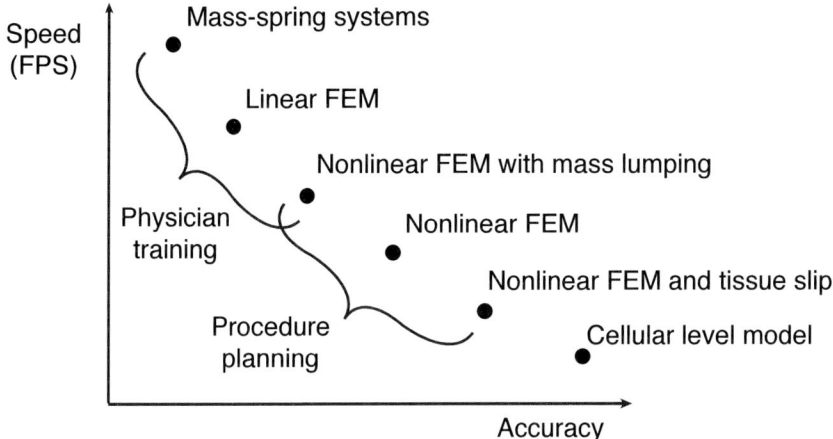

Fig. 8.2. There is an inherent trade-off between the accuracy and speed (as measured in frames-per-second) of physically-based simulation algorithms for medical procedures involving soft tissue deformations. While simulation for physician training is subject to strict real-time performance requirements, accuracy is more critical for patient-specific procedure planning. New algorithms can push the trade-off frontier outwards, enabling faster, more accurate simulations.

model tissue dynamics, another important aspect is to explicitly model slip between tissue types. Rather than considering the mesh of heterogeneous tissue to be connected, the simulation should allow each independently meshed tissue type to deform and move independently and affect neighboring tissues through collision and frictional forces. For computational efficiency, constructing oct-trees and related data structures may accelerate collision detection of deformable objects [99].

Finally, it is crucial to validate these new simulation and planning algorithms using data from physical experiments. Testbeds using artificial tissue phantoms can provide detailed information about deformations, including the precise displacements of points inside deforming tissues [119] and the forces that occur during needle insertion [70]. The needle steering testbed being developed at Johns Hopkins University, shown in figure 8.3, provides an ideal platform for obtaining precise deformation and force measurements for steerable needles in deforming tissue [208, 209]. Ultimately, simulations will also need to be validated using medical images from animal studies or patient cases.

8.2.2 Planning Algorithms for Image-Guided Medical Procedures

New treatment technologies also introduce new computational planning challenges. In this book, we discussed motion planning for steerable needles, a new treatment approach which was submitted in 2006 for a United States Patent [211].

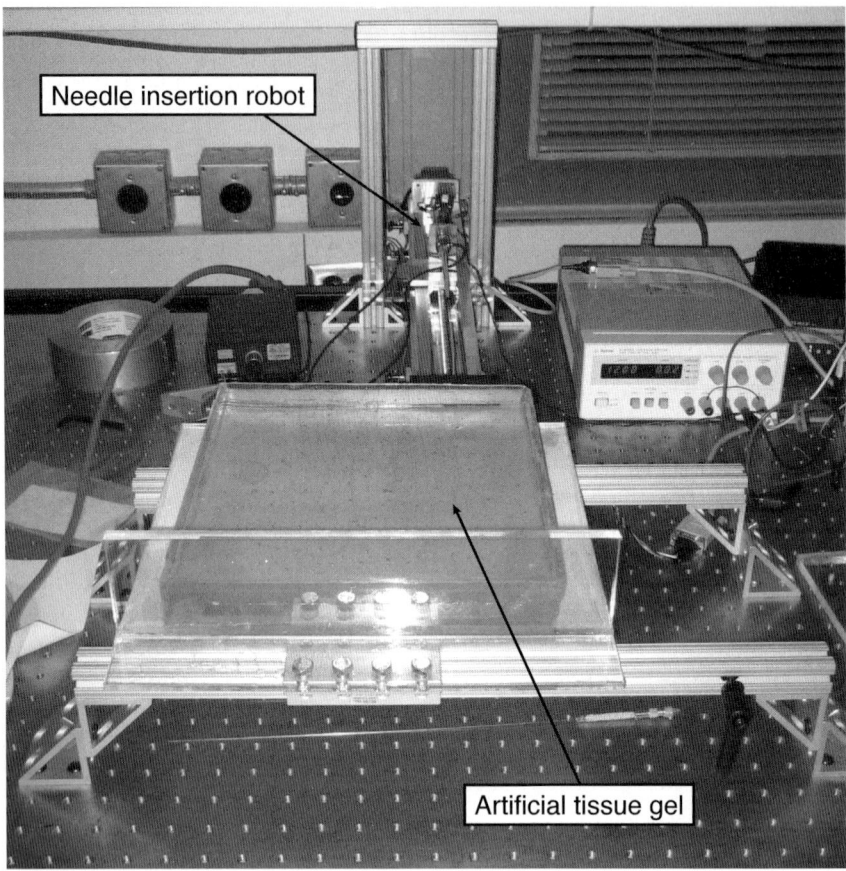

Fig. 8.3. The needle steering testbed at Johns Hopkins University [208, 209] includes a robotic device that can insert and steer a needle in a semi-transparent artificial tissue gel. Optical cameras mounted above the testbed (not shown) can precisely track the motion of the needle and deformations of the artificial tissue gel, enabling physical validation of computational simulations and motion planning algorithms for steerable needles.

New extensions to needle steering are already being proposed that change the dynamic behavior of the needle, including a flattened and expanded bevel tip [76]. In future work, we will explore motion planning for steerable needles in 3-D volumes, where needles nominally follow helical trajectories.

Recently there has been a renaissance in research on thermal therapy for cancer treatment, where directional beams of heat are used to kill cancer cells. This type of therapy raises dose planning problems mathematically similar to HDR brachytherapy dose optimization but can offer a new capability: directional beams [66]. This treatment approach also introduces new challenges for planning since the plan must be updated in real-time during treatment to override the natural tendency of living tissues to compensate for temperature changes via blood flow.

Another open area for research is combining motion planning with uncertainty and deformations. Because patient-specific tissue parameters generally cannot be known exactly, we would like to develop probabilistic models to compute and display uncertainty in tissue deformations. Developing new classes of motion planners that explicitly consider motion uncertainty and deformations and their combined effect on the planning objective will improve the effectiveness of operating inside the highly variable environment of living tissue.

8.2.3   New Clinical Applications

Although this monograph focuses on the specific application of prostate brachytherapy, we believe these motion planning algorithms can be applied to many other medical procedures, including anesthesia injections, biopsies, and cryotherapy. Needle steering could be performed in a variety of soft tissue types, such as inside the liver or brain. Future work is needed to expand the applicability of the methods presented in this monograph; this will require physician feedback and adaptation for the nuances of each procedure and tissue type.

8.3   Conclusion

In medicine, improvements in imaging, combined with new devices to reach what is revealed, can lead to improved treatment and patient health. Innovations in medical imaging are constantly introducing new and improved imaging modalities. Mega-voltage Cone-beam Computed Tomography (MVCBCT) is a new imaging modality that uses the same linac as Intensity Modulated Radiation Therapy (IMRT), allowing physicians to image and treat the patient using the same device without moving and re-aligning the patient [159, 175]. Optical coherence tomography (OCT) is an optical *in vivo* imaging method with micron-scale resolution that can be used to visualize the tiny intra-retinal and intra-corneal anatomy [73]. Molecular imaging techniques are enabling radiologists to visualize *in vivo* intracellular biochemical pathways, introducing the potential to identify pathways involved in disease before traditional symptoms appear [212]. Each new imaging modality introduces a wealth of new digital information, which introduces new computational challenges. In parallel, new robotic medical devices are being introduced, from needle insertion robots [83, 169, 172, 209] to active cannulas capable of following curved paths in soft tissue as well as open cavities [210] to magnetically controlled micro-robots operating in the eye [78]. By bridging the gap between medical imaging and robotic surgical devices with motion planning algorithms, our aim is to assist physicians and thus improve patient care.

References

1. Aarts, E., Korst, J.: Simulated Annealing and Boltzmann Machines, 1st edn. Interscience Series in Discrete Mathematics and Optimization. John Wiley & Sons, Inc. (1989)
2. Abolhassani, N., Patel, R.V., Ayazi, F.: Minimization of needle deflection in robot-assisted percutaneous therapy. Int. J. Medical Robotics and Computer Assisted Surgery 3(2), 140–148 (2007)
3. Abolhassani, N., Patel, R.V., Moallem, M.: Needle insertion into soft tissue: a survey. Medical Engineering & Physics 29(4), 413–431 (2007)
4. Agarwal, P.K., Biedl, T., Lazard, S., Robbins, S., Suri, S., Whitesides, S.: Curvature-constrained shortest paths in a convex polygon. SIAM J. Comput. 31(6), 1814–1851 (2002)
5. Akella, S., Huang, W., Lynch, K.M., Mason, M.T.: Sensorless parts feeding with a one joint robot. In: Laumond, J.-P., Overmars, M. (eds.) Algorithms for Robotic Motion and Manipulation: 1996 WAFR, Wellesley, MA, pp. 229–237. AK Peters, Ltd. (1997)
6. Alagoz, O., Maillart, L.M., Schaefer, A.J., Roberts, M.: The optimal timing of living-donor liver transplantation. Management Science 50(10), 1420–1430 (2005)
7. Alterovitz, R., Branicky, M., Goldberg, K.: Constant curvature motion planning under uncertainty with applications in image-guided medical needle steering. In: Proc. Int. Workshop on the Algorithmic Foundations of Robotics (July 2006)
8. Alterovitz, R., Goldberg, K., Kurhanewicz, J., Pouliot, J., Hsu, I.-C.: Image registration for prostate MR spectroscopy using biomechanical modeling and optimization of force and stiffness parameters. In: Proc. Int. Conf. IEEE Engineering In Medicine and Biology Society (September 2004)
9. Alterovitz, R., Goldberg, K., Kurhanewicz, J., Pouliot, J., Hsu, I.-C.: Registering MR with MRS images for HDR prostate treatment using finite element modeling. In: American Association of Physicists in Medicine (AAPM) 46th Annual Meeting (July 2004)
10. Alterovitz, R., Goldberg, K., Okamura, A.M.: Planning for steerable bevel-tip needle insertion through 2D soft tissue with obstacles. In: Proc. IEEE Int. Conf. Robotics and Automation (ICRA), pp. 1652–1657 (April 2005)
11. Alterovitz, R., Goldberg, K., Pouliot, J., Hsu, I.-C.J., Kim, Y., Noworolski, S.M., Kurhanewicz, J.: Registration of MR prostate images with biomechanical modeling and nonlinear parameter estimation. Med. Phys. 33(2), 446–454 (2006)

12. Alterovitz, R., Kim, Y., Kurhanewicz, J., Pouliot, J., Hsu, I.-C.J., Goldberg, K.: Prostate MR spectroscopy image registration using biomechanical modeling of tissue deformations due to endorectal probe insertion. In: Annual Meeting of the American Brachytherapy Society (June 2005)

13. Alterovitz, R., Lessard, E., Pouliot, J., Hsu, I.-C.J., O'Brien, J.F., Goldberg, K.: High-dose-rate brachytherapy dose optimization for prostate cancer using linear programming. In: Annual Meeting of the Institute for Operations Research and the Management Sciences (INFORMS) (November 2005)

14. Alterovitz, R., Lessard, E., Pouliot, J., Hsu, I.-C.J., O'Brien, J.F., Goldberg, K.: Optimization of HDR brachytherapy dose distributions using linear programming with penalty costs. Med. Phys. 33(11), 4012–4019 (2006)

15. Alterovitz, R., Lim, A., Goldberg, K., Chirikjian, G.S., Okamura, A.M.: Steering flexible needles under Markov motion uncertainty. In: Proc. IEEE/RSJ Int. Conf. on Intelligent Robots and Systems (IROS), pp. 120–125 (August 2005)

16. Alterovitz, R., Pouliot, J., Taschereau, R., Hsu, I.-C., Goldberg, K.: Modeling seed misplacement by simulating tissue deformations. In: Annual Meeting of the American Brachytherapy Society (May 2003)

17. Alterovitz, R., Pouliot, J., Taschereau, R., Hsu, I.-C., Goldberg, K.: Needle insertion and radioactive seed implantation in human tissues: Simulation and sensitivity analysis. In: Proc. IEEE Int. Conf. Robotics and Automation (ICRA), vol. 2, pp. 1793–1799 (September 2003)

18. Alterovitz, R., Pouliot, J., Taschereau, R., Hsu, I.-C., Goldberg, K.: Sensorless planning for medical needle insertion procedures. In: Proc. IEEE/RSJ Int. Conf. on Intelligent Robots and Systems (IROS), vol. 3, pp. 3337–3343 (October 2003)

19. Alterovitz, R., Pouliot, J., Taschereau, R., Hsu, I.-C., Goldberg, K.: Simulating needle insertion and radioactive seed implantation for prostate brachytherapy. In: Westwood, J.D., et al. (eds.) Medicine Meets Virtual Reality 11, Washington, DC, pp. 19–25. IOS Press, Amsterdam (2003)

20. Amato, N.M., Bayazit, O.B., Dale, L.K., Jones, C., Vallejo, D.: OBPRM: An obstacle-based PRM for 3D workspaces. In: Agarwal, P., Kavraki, L.E., Mason, M. (eds.) Robotics: The Algorithmic Perspective: 1998 WAFR, Natick, MA, pp. 156–168. AK Peters, Ltd. (1998)

21. Apaydin, M., Brutlag, D., Guestrin, C., Hsu, D., Latombe, J.-C.: Stochastic conformational roadmaps for computing ensemble properties of molecular motion. In: Boissonnat, J.-D., Burdick, J., Goldberg, K., Hutchinson, S. (eds.) Algorithmic Foundations of Robotics V, Berlin, Germany, pp. 131–148. Springer, Heidelberg (2003)

22. Apaydin, M.S., Brutlag, D.L., Guestrin, C., Hsu, D., Latombe, J.-C.: Stochastic roadmap simulation: an efficient representation and algorithm for analyzing molecular motion. In: Proc. RECOMB, pp. 12–21 (2002)

23. Atramentov, A., LaValle, S.M.: Efficient nearest neighbor searching for motion planning. In: Proc. IEEE Int. Conf. Robotics and Automation (ICRA) (2002)

24. Audette, M.A., Delingette, H., Fuchs, A., Astley, O.R., Chinzei, K.: A topologically faithful, tissue-guided, spatially varying meshing strategy for computing patient-specific head models for endoscopic pituitary surgery simulation. In: Westwood, J.D., et al. (eds.) Medicine Meets Virtual Reality 14, Washington, DC, pp. 22–27. IOS Press, Amsterdam (2005)

25. Azar, F.S., Metaxas, D.N., Schnall, M.D.: A deformable finite element model of the breast for predicting mechanical deformations under external perturbations. Academic Radiology 8(10), 965–975 (2001)

26. Bajcsy, R., Broit, C.: Matching of deformed images. In: 6th Int. Conf. on Pattern Recognition, pp. 351–353 (1982)

27. Bakker, B., Zivkovic, Z., Krose, B.: Hierarchical dynamic programming for robot path planning. In: Proc. IEEE/RSJ Int. Conf. on Intelligent Robots and Systems (IROS), pp. 2756–2761 (August 2005)

28. Baraff, D., Witkin, A.: Large steps in cloth simulation. Computer Graphics (Proc. SIGGRAPH 1998) 32, 43–54 (1998)

29. Barker, J.L., Garden, A., Ang, K.K., O'Daniel, J.C., Wang, H., Court, L.E., Morrison, W.H., Rosenthal, D., Chao, K., Tucker, S., Mohan, R., Dong, L.: Quantification of volumetric and geometric changes occurring during fractionated radiotherapy for head-and-neck cancer using an integrated CT/linear accelerator system. Int. J. Radiat. Oncol. Biol. Phys. 59(4), 960–970 (2004)

30. Barto, A., Bradtke, S., Singh, S.: Learning to act using real-time dynamic programming. Artificial Intelligence 72(1-2), 81–138 (1995)

31. Bazaraa, M.S., Sherali, H.D., Shetty, C.M.: Nonlinear Programming: Theory and Algorithms, 2nd edn. John Wiley & Sons, Inc., New York (1993)

32. Bemporad, A., Morari, M.: Control of systems integrating logic, dynamics and constraints. Automatica 35(3), 407–427 (1999)

33. Ben-Tal, A., Nemirovski, A.: Robust solutions of uncertain linear programs. Operations Research Letters 25(1), 1–13 (1999)

34. Bertsekas, D.P.: Dynamic Programming and Optimal Control, 2nd edn. Athena Scientific, Belmont (2000)

35. Bharatha, A., Hirose, M., Hata, N., Warfield, S.K., Ferrant, M., Zou, K.H., Suarez-Santana, E., Ruiz-Alzola, J., D'Amico, A., Cormack, R.A., Kikinis, R., Jolesz, F.A., Tempanya, C.M.C.: Evaluation of three-dimensional finite element-based deformable registration of pre- and intraoperative prostate imaging. Med. Phys. 28(12), 2551–2560 (2001)

36. Bicchi, A., Casalino, G., Santilli, C.: Planning shortest bounded-curvature paths for a class of nonholonomic vehicles among obstacles. J. Intelligent and Robotic Systems 16, 387–405 (1996)

37. Böhringer, K.-F., Bhatt, V., Donald, B.R., Goldberg, K.: Algorithms for sensorless manipulation using a vibrating surface. Algorithmica 26(3/4), 389–429 (2000)

38. Boor, V., Overmars, N.H., van der Stappen, A.F.: The Gaussian sampling strategy for probabilisitic roadmap planners. In: Proc. IEEE Int. Conf. Robotics and Automation (ICRA), pp. 1018–1023 (1999)

39. Bouilly, B., Siméon, T., Alami, R.: A numerical technique for planning motion strategies of a mobile robot in presence of uncertainty. In: Proc. IEEE Int. Conf. Robotics and Automation (ICRA), vol. 2, pp. 1327–1332 (May 1995)

40. Branicky, M., Borkar, V., Mitter, S.: A unified framework for hybrid control: Model and optimal control theory. IEEE Trans. Automatic Control 43(1), 31–45 (1998)

41. Branicky, M., Hebbar, R., Zhang, G.: A fast marching algorithm for hybrid systems. In: Proc. IEEE Conf. on Decision and Control, pp. 4897–4902 (December 1999)

42. Bro-Nielsen, M., Tasto, J., Cunningham, R., Merril, G.: PreOp endoscopic simulator: A PC-based immersive training system. In: Westwood, J.D., et al. (eds.) Medicine Meets Virtual Reality, pp. 76–82. IOS Press, Amsterdam (1999)

43. Brost, R.C.: Automatic grasp planning in the presence of uncertainty. Int. J. Robotics Research 7(1), 3–17 (1988)

44. Brouwer, I., Ustin, J., Bentley, L., Sherman, A., Dhruv, N., Tendick, F.: Measuring in vivo animal soft tissue properties for haptic modeling in surgical simulation. In: Westwood, J.D., et al. (eds.) Medicine Meets Virtual Reality, pp. 69–74. IOS Press, Amsterdam (2001)

45. Brown, J., Sorkin, S., Bruyns, C., Latombe, J., Montgomery, K., Stephanides, M.: Real-time simulation of deformable objects: tools and applications. Computer Animation (November 2001)

46. Brown, J.D., Rosen, J., Kim, Y.S., Chang, L., Sinanan, M.N., Hannaford, B.: In-vivo and in-situ compressive properties of porcine abdominal soft tissues. In: Westwood, J.D., et al. (eds.) Medicine Meets Virtual Reality 11, pp. 26–32. IOS Press, Amsterdam (2003)

47. Burns, B., Brock, O.: Toward optimal configuration space sampling. In: Proc. Robotics: Science and Systems, Cambridge, MA (June 2005)

48. Cahill, D.R., Orland, M.J., Miller, G.M.: Atlas of human cross-sectional anatomy: with CT and MR Images, 3rd edn. Wiley-Liss, New York (1995)

49. Cavusoglu, M.C.: Telesurgery and Surgical Simulation: Design, Modeling and Evaluation of Haptic Interfaces to Real and Virtual Surgical Environments. PhD thesis, University of California, Berkeley (2000)

50. Chinzei, K., Hata, N., Jolesz, F.A., Kikinis, R.: MR compatible surgical assist robot: System integration and preliminary feasibility study. In: Delp, S.L., DiGoia, A.M., Jaramaz, B. (eds.) MICCAI 2000. LNCS, vol. 1935, pp. 921–930. Springer, Heidelberg (2000)

51. Choset, H., Lynch, K.M., Hutchinson, S., Kantor, G., Burgard, W., Kavraki, L.E., Thrun, S.: Principles of Robot Motion: Theory, Algorithms, and Implementations. MIT Press, Cambridge (2005)

52. Chow, C.-S., Tsitsiklis, J.: An optimal one-way multigrid algorithm for discrete-time stochastic control. IEEE Trans. Automatic Control 36(8), 898–914 (1991)

53. Christensen, G., Miller, M., Vannier, M., Grenander, U.: Individualizing neuroanatomical atlases using a massively parallel computer. IEEE Computer 29(1), 32–38 (1996)

54. Citrin, D., Ning, H., Guion, H., Li, G., Susil, R., Miller, R., Lessard, E., Pouliot, J., Huchen, X., Capala, J., Coleman, C., Camphausen, K., Menard, C.: Inverse treatment planning based on MRI for HDR prostate brachytherapy. Int. J. Radiat. Oncol. Biol. Phys. 61(4), 1267–1275 (2005)

55. Cleary, K., Ibanez, L., Navab, N., Stoianovici, D., Patriciu, A., Corral, G.: Segmentation of surgical needles for fluoroscopy servoing using the Insight Software Toolkit (ITK). In: Proc. Int. Conf. IEEE Engineering In Medicine and Biology Society, pp. 698–701 (September 2003)

56. Cleary, K., Melzer, A., Watson, V., Kronreif, G., Stoianovici, D.: Interventional robotic systems: applications and technology state-of-the-art. Minimally Invasive Therapy & Allied Technologies 15(2), 101–113 (2006)

57. Cotin, S., Delingette, H.: Real-time surgery simulation with haptic feedback using finite elements. In: Proc. IEEE Int. Conf. Robotics and Automation (ICRA), pp. 3739–3744 (1998)

58. Cotin, S., Delingette, H., Ayache, N.: Real-time elastic deformations of soft tissues for surgery simulation. IEEE Trans. Visualization and Computer Graphics 5(1), 62–73 (1999)

59. Crouch, J.R.: Medial Techniques for Automating Finite Element Analysis. PhD thesis, University of North Carolina, Chapel Hill (2003)

60. Crouch, J.R., Pizer, S.M., Chaney, E.L., Zaider, M.: Medially based meshing with finite element analysis of prostate deformation. In: Ellis, R.E., Peters, T.M. (eds.) MICCAI 2003. LNCS, vol. 2878, pp. 108–115. Springer, Heidelberg (2003)

61. Crouch, J.R., Schneider, C.M., Wainer, J., Okamura, A.M.: A velocity-dependent model for needle insertion in soft tissue. In: Duncan, J.S., Gerig, G. (eds.) MICCAI 2005. LNCS, vol. 3750, pp. 624–632. Springer, Heidelberg (2005)

62. Dawson, J.E., Wu, T., Roy, T., Gy, J.Y., Kim, H.: Dose effects of seeds placement deviations from pre-planned positions in ultrasound guided prostate implants. Radiotherapy and Oncology 32(2), 268–270 (1994)

63. Dean, T., Kaelbling, L.P., Kirman, J., Nicholson, A.: Planning under time constraints in stochastic domains. Artificial Intelligence 76(1-2), 35–74 (1995)

64. DeWitt, K., Hsu, I.-C., Weinberg, V., Lessard, E., Pouliot, J.: 3-d inverse treatment planning for the tandem and ovoid applicator in cervical cancer. Int. J. Radiat. Oncol. Biol. Phys. 63(4), 1270–1274 (2005)

65. Dice, L.R.: Measures of the amount of ecologic association between species. Ecology 26(3), 297–302 (1945)

66. Diederich, C.J., Stafford, R.J., Nau, W.H., Burdette, E.C., Price, R.E., Hazle, J.D.: Transurethral ultrasound applicators with directional heating patterns for prostate thermal therapy: In vivo evaluation using MR thermometry. Medical Physics 31(2), 405–413 (2004)

67. DiMaio, S.P., Kacher, D.F., Ellis, R.F., Fichtinger, G., Hata, N., Zientara, G.P., Panych, L.P., Kikinis, R., Jolesz, F.A.: Needle artifact localization in 3T MR images. In: Westwood, J.D., et al. (eds.) Medicine Meets Virtual Reality 14, Washington, DC, pp. 120–125. IOS Press, Amsterdam (2006)

68. DiMaio, S.P., Pieper, S., Chinzei, K., Hata, N., Balogh, E., Fichtinger, G., Tempany, C.M., Kikinis, R.: Robot-assisted needle placement in open-MRI: System architecture, integration and validation. In: Westwood, J.D., et al. (eds.) Medicine Meets Virtual Reality 14, Washington, DC, pp. 126–131. IOS Press, Amsterdam (2006)

69. DiMaio, S.P., Salcudean, S.E.: Needle insertion modelling and simulation. In: Proc. IEEE Int. Conf. Robotics and Automation (ICRA), pp. 2098–2105 (May 2002)

70. DiMaio, S.P., Salcudean, S.E.: Needle insertion modelling and simulation. IEEE Trans. Robotics and Automation 19(5), 864–875 (2003)

71. DiMaio, S.P., Salcudean, S.E.: Needle steering and model-based trajectory planning. In: Ellis, R.E., Peters, T.M. (eds.) MICCAI 2003. LNCS, vol. 2878, pp. 33–40. Springer, Heidelberg (2003)

72. DiMaio, S.P., Salcudean, S.E.: Needle steering and motion planning in soft tissues. IEEE Trans. Biomedical Engineering 52(6), 965–974 (2005)

73. Drexler, W., Morgner, U., Ghanta, R.K., Kärtner, F.X., Schuman, J.S., Fujimoto, J.G.: Ultrahigh-resolution ophthalmic optical coherence tomography. Nature Medicine 7, 502–507 (2001)

74. Dubins, L.E.: On curves of minimal length with a constraint on average curvature and with prescribed initial and terminal positions and tangents. American J. of Mathematics 79(3), 497–516 (1957)

75. Duindam, V., Alterovitz, R., Sastry, S., Goldberg, K.: Screw-based motion planning for bevel-tip flexible needles in 3d environments with obstacles. In: Proc. IEEE Int. Conf. Robotics and Automation (ICRA) (May 2008)

76. Engh, J., Podnar, G., Kondziolka, D., Riviere, C.: Toward effective needle steering in brain tissue. In: Proc. Int. Conf. IEEE Engineering In Medicine and Biology Society, pp. 559–562 (September 2006)

77. Erdmann, M., Mason, M.: An exploration of sensorless manipulation. IEEE J. Robotics and Automation 4(4), 369–379 (1988)

78. Ergeneman, O., Dogangil, G., Kummer, M.P., Jake, M., Abbott, J., Nazeeruddin, M.K., Nelson, B.J.: A magnetically controlled wireless optical oxygen sensor for intraocular measurements. IEEE Sensors Journal 8(1), 29–37 (2008)

79. Fei, B., Kemper, C., Wilson, D.L.: A comparative study of warping and rigid body registration for the prostate and pelvic MR volumes. Computerized Medical Imaging and Graphics 27, 267–281 (2003)

80. Fei, B., Wheaton, A., Lee, Z., Duerk, J.I., Wilson, D.: Automatic MR volume registration and its evaluation for the pelvis and prostate. Physics in Medicine and Biology 47(5), 823–838 (1999)

81. Ferguson, D., Stentz, A.: Focussed dynamic programming: Extensive comparative results. Technical Report CMU-RI-TR-04-13, Robotics Institute, Carnegie Mellon University, Pittsburgh, PA (March 2004)

82. Ferguson, D., Stentz, A.: Focussed processing of MDPs for path planning. In: Proc. IEEE International Conference on Tools with Artificial Intelligence (IC-TAI), November 2004, pp. 310–317 (2004)

83. Fichtinger, G., Burdette, E.C., Tanacs, A., Patriciu, A., Mazilu, D., Whitcomb, L.L., Stoianovici, D.: Robotically assisted prostate brachytherapy with transrectal ultrasound guidance–phantom experiments. Brachytherapy 5(1), 14–26 (2006)

84. Fichtinger, G., DeWeese, T.L., Patriciu, A., Tanacs, A., Mazilu, D., Anderson, J.H., Masamune, K., Taylor, R.H., Stoianovici, D.: System for robotically assisted prostate biopsy and therapy with intraoperative CT guidance. Academic Radiology 9(1), 60–74 (2002)

85. Fourer, R., Gay, D.M., Kernighan, B.W.: AMPL: a modeling language for mathematical programming, 2nd edn. Thomson/Brooks/Cole, Pacific Grove (2003)

86. Fung, Y.C.: Biomechanics: Mechanical Properties of Living Tissues, 2nd edn. Springer, Heidelberg (1993)

87. Fung, Y.C.: A First Course in Continuum Mechanics, 3rd edn. Prentice Hall, Englewood Cliffs (1994)

88. Gallagher, A.G., Ritter, E.M., Champion, H., Higgins, G., Fried, M.P., Moses, G., Smith, C.D., Satava, R.M.: Virtual reality simulation for the operating room: proficiency-based training as a paradigm shift in surgical skills training. Annals of Surgery 241(2), 364–372 (2005)

89. Geman, S., Geman, D.: Stochastic relaxation, Gibbs distribution, and the Bayesian restoration of images. IEEE Trans. Patt. Anal. Mac. Int. 6, 721–741 (1984)

90. General Electric Company. PROSE Pulse Sequence, ch.6 (2002)

91. Gibson, S.F., Mirtich, B.: A survey of deformable modeling in computer graphics. Technical report, MERL, TR-97-19 (1997)

92. Glozman, D., Shoham, M.: Image-guided robotic flexible needle steering. IEEE Trans. Robotics 23(3), 459–467 (2007)

93. Goksel, O., Salcudean, S.E., DiMaio, S.P.: 3d simulation of needle-tissue interaction with application to prostate brachytherapy. Computer Aided Surgery 11(6), 279–288 (2006)

94. Goldberg, K.Y.: Orienting polygonal parts without sensors. Algorithmica 10(3), 201–255 (1993)

95. Hansen, E.A., Zilberstein, S.: LAO*: A heuristic search algorithm that finds solutions with loops. Artificial Intelligence 129(1), 35–62 (2001)

96. Hansen, E.K., Bucci, K., Quivey, J.M., Weinberg, V., Xia, P.: Repeat CT imaging and re-planning during the course of IMRT for head and neck cancer. Int. J. Radiat. Oncol. Biol. Phys. 64(2), 355–362 (2006)

97. Hata, N., Hashimoto, R., Tokuda, J., Morikawa, S.: Needle guiding robot for MR-guided microwave thermotherapy of liver tumor using motorized remote-center-of-motion constraint. In: Proc. IEEE Int. Conf. Robotics and Automation (ICRA), April 2005, pp. 1652–1656 (2005)

98. Hayes, C.E., Dietz, M.J., King, B.F., Ehman, R.L.: Pelvic imaging with phased array coils: quantitative assessment of signal-to-noise ratio improvement. J. Magnetic Resonance Imaging 2(3), 321–326 (1992)

99. Hearn, D., Baker, M.P.: Computer Graphics with OpenGL, 3rd edn. Prentice Hall, Englewood Cliffs (2003)

100. Heverly, M., Dupont, P., Triedman, J.: Trajectory optimization for dynamic needle insertion. In: Proc. IEEE Int. Conf. Robotics and Automation (ICRA), April 2005, pp. 1646–1651 (2005)

101. Hing, J.T., Brooks, A.D., Desai, J.P.: A biplanar fluoroscopic approach for the measurement, modeling, and simulation of needle and soft-tissue interaction. Medical Image Analysis 11(1), 62–78 (2007)

102. Horn, B.K.P.: Closed-form solution of absolute orientation using unit quaternions. J. Optical Society of America A 4(4), 629–642 (1987)

103. Howe, R.D., Matsuoka, Y.: Robotics for surgery. Annual Review of Biomedical Engineering 1, 211–240 (1999)

104. Hricak, H., White, S., Vigneron, D., Kurhanewicz, J., Kosco, A., Levin, D., Weiss, J., Narayan, P., Carroll, P.: Carcinoma of the prostate gland: MR imaging with pelvic phased-array coils versus integrated endorectal–pelvic phased-array coils. Radiology 193(3), 703–709 (1994)

105. Hsiao, K., Kaelbling, L.P., Lozano-Perez, T.: Grasping POMDPs. In: Proc. IEEE Int. Conf. Robotics and Automation (ICRA), April 2007, pp. 4685–4692 (2007)

106. Hsu, I.-C.J., Lessard, E., Weinberg, V., Pouliot, J.: Comparison of inverse planning simulated annealing and geometrical optimization for prostate high-dose-rate brachytherapy. Brachytherapy 3(3), 147–152 (2004)

107. Hui, S., Das, R.: Optimization of conformal avoidance: a comparative study of prone vs. supine interstitial high-dose-rate breast brachytherapy. Brachytherapy 4(2), 137–140 (2005)

108. Husband, J.E., Padhani, A.R., MacVicar, A.D., Revell, P.: Magnetic resonance imaging of prostate cancer: comparison of image quality using endorectal and pelvic phased array coils. Clinical Radiology 53(9), 673–681 (1998)

109. ILOG, Inc. ILOG CPLEX: High-performance software for mathematical programming and optimization (2005), http://www.ilog.com/products/cplex/

110. Jacob, R., Hanlon, A., Horwitz, E., Movsas, B., Uzzo, R., Pollack, A.: The relationship of increasing radiotherapy dose to reduced distant metastases and mortality in men with prostate cancer. Cancer 100(3), 538–543 (2004)

111. Jacobs, P., Canny, J.: Planning smooth paths for mobile robots. In: Proc. IEEE Int. Conf. Robotics and Automation (ICRA), May 1989, pp. 2–7 (1989)

112. James, D.L., Pai, D.K.: ARTDEFO: Accurate real time deformable objects. Computer Graphics (Proc. SIGGRAPH 1999), 65–72 (1999)

113. Jemal, A., Tiwari, R.C., Murray, T., Ghafoor, A., Samuels, A., Ward, E., Feuer, E.J., Thun, M.J.: Cancer statistics, 2004. CA: A Cancer Journal for Clinicians 54, 8–29 (2004)

114. Jozsef, G., Streeter, O.E., Astrahan, M.A.: The use of linear programming in optimization of HDR implant dose distributions. Med. Phys. 30(5), 751–760 (2003)

115. Kalanovic, D., Ottensmeyer, M.P., Gross, J., Buess, G., Dawson, S.L.: Independent testing of soft tissue visco-elasticity using indentation and rotary shear deformations. In: Westwood, J.D., et al. (eds.) Medicine Meets Virtual Reality 11, January 2003, pp. 137–143. IOS Press, Amsterdam (2003)

116. Kallem, V., Cowan, N.J.: Image-guided control of flexible bevel-tip needles. In: Proc. IEEE Int. Conf. Robotics and Automation (ICRA) (2007)

117. Kataoka, H., Washio, T., Chinzei, K., Mizuhara, K., Simone, C., Okamura, A.: Measurement of tip and friction force acting on a needle during penetration. In: Dohi, T., Kikinis, R. (eds.) MICCAI 2002. LNCS, vol. 2488, pp. 216–223. Springer, Heidelberg (2002)

118. Kavraki, L.E., Svestka, P., Latombe, J.-C., Overmars, M.: Probabilistic roadmaps for path planning in high dimensional configuration spaces. IEEE Trans. Robotics and Automation 12(4), 566–580 (1996)

119. Kerdok, A.E., Cotin, S.M., Ottensmeyer, M.P., Galeaa, A.M., Howe, R.D., Dawson, S.L.: Truth cube: Establishing physical standards for soft tissue simulation. Medical Image Analysis 7, 283–291 (2003)

120. Kessler, M., Roberson, M., Zeng, R., Fessler, J.: Deformable image registration using multiresolution B-splines. Medical Physics (Abstract) 31, 1792 (2004)

121. Kim, Y.: Image-Based High Dose Rate (HDR) Brachytherapy for Prostate Cancer. PhD thesis, University of California, Berkeley (December 2003)

122. Kim, Y., Noworolski, S.M., Pouliot, J., Hsu, I.-C.J., Vigneron, D.B., Kurhanewicz, J.: Expandable and rigid endorectal coils for prostate MRI: Impact on prostate distortion and rigid image registration. Med. Phys. 32(12), 3569–3578 (2005)

123. Kneschaurek, P., Schiess, W., Wehrmann, R.: Volume-based dose optimization in brachytherapy. Int. J. Radiat. Oncol. Biol. Phys. 45(3), 811–815 (1999)

124. Kobayashi, Y., Okamoto, J., Fujie, M.: Physical properties of the liver and the development of an intelligent manipulator for needle insertion. In: Proc. IEEE Int. Conf. Robotics and Automation (ICRA), April 2005, pp. 1632–1639 (2005)

125. Kreke, J., Schaefer, A.J., Roberts, M., Bailey, M.: Optimizing testing and discharge decisions in the management of severe sepsis. In: Annual Meeting of INFORMS (November 2005)

126. Krouskop, T.A., Wheeler, T.M., Kallel, F., Garria, B.S., Hall, T.: Elastic moduli of breast and prostate tissues under compression. Ultrasonic Imaging 20(4), 260–274 (1998)

127. Kuban, D., Pollack, A., Huang, E., Levy, L., Dong, L., Starkschall, G., Rosen, I.: Hazards of dose escalation in prostate cancer radiotherapy. Int. J. Radiat. Oncol. Biol. Phys. 57(5), 1260–1268 (2003)

128. Kurhanewicz, J., Swanson, M., Nelson, S.J., Vigneron, D.: Combined magnetic resonance imaging and spectroscopic imaging approach to molecular imaging of prostate cancer. J. Magnetic Resonance Imaging 16(4), 451–463 (2002)

129. Kurhanewicz, J., Vigneron, D., Nelson, S.J.: Three-dimensional magnetic resonance spectroscopic imaging of brain and prostate cancer. Neoplasia 2(1-2), 166–189 (2000)

130. Kurhanewicz, J., Vigneron, D.B., Hricak, H., Narayan, P., Carroll, P., Nelson, S.J.: Three-dimensional H-1 MR spectroscopic imaging of the in situ human prostate with high (0.24-0.7-cm3) spatial resolution. Radiology 198(3), 795–805 (1996)

131. Kurhanewicz, J., Vigneron, D.B., Hricak, H., Parivar, F., Nelson, S.J., Shinohara, K., Carroll, P.R.: Prostate cancer: Metabolic response to cryosurgery as detected with 3D H-1 MR spectroscopic imaging. Radiology 200(2), 489–496 (1996)

132. Lachance, B., Beliveau-Nadeau, D., Lessard, E., Chretien, M., Hsu, I.C., Pouliot, J., Beaulieu, L., Vigneault, E.: Early clinical experience with anatomy-based inverse planning dose optimization for high-dose-rate boost of the prostate. Int. J. Radiat. Oncol. Biol. Phys. 54(1), 86–100 (2002)

133. Lahanas, M., Baltas, D., Giannouli, S., Milickovic, N., Zamboglou, N.: Generation of uniformly distributed dose points for anatomy-based three-dimensional dose optimization methods in brachytherapy. Med. Phys. 27(5), 1034–1046 (2000)

134. Lahanas, M., Baltas, D., Zamboglou, N.: Anatomy-based three dimensional dose optimization in brachytherapy using multiobjective genetic algorithms. Med. Phys. 26(9), 1904–1918 (1999)

135. Lahanas, M., Baltas, D., Zamboglou, N.: A hybrid evolutionary algorithm for multi-objective anatomy-based dose optimization in high dose rate brachytherapy. Physics in Medicine and Biology 48(3), 399–415 (2003)

136. Latombe, J.-C.: Robot Motion Planning. Kluwer Academic Pub., Dordrecht (1991)

137. Latombe, J.-C.: Motion planning: A journey of robots, molecules, digital actors, and other artifacts. Int. J. Robotics Research 18(11), 1119–1128 (1999)

138. Laumond, J.-P., Jacobs, P.E., Taïx, M., Murray, R.M.: A motion planner for nonholonomic mobile robots. IEEE Trans. Robotics and Automation 10(5), 577–593 (1994)

139. LaValle, S.M.: Planning Algorithms. Cambridge University Press, Cambridge (2006)

140. LaValle, S.M., Branicky, M.S., Lindemann, S.R.: On the relationship between classical grid search and probabilistic roadmaps. Int. J. Robotics Research 23(7/8), 673–692 (2004)

141. LaValle, S.M., Hutchinson, S.A.: An objective-based framework for motion planning under sensing and control uncertainties. Int. J. Robotics Research 17(1), 19–42 (1998)

142. LaValle, S.M., Lin, D., Guibas, L.J., Latombe, J.-C., Motwani, R.: Finding an unpredictable target in a workspace with obstacles. In: Proc. IEEE Int. Conf. Robotics and Automation (ICRA), pp. 737–742 (1997)

143. Lazanas, A., Latombe, J.: Motion planning with uncertainty: A landmark approach. Artificial Intelligence 76(1-2), 285–317 (1995)

144. Lessard, E.: Development and clinical introduction of an inverse planning dose optimization by simulated annealing (IPSA) for high dose rate brachytherapy. Medical Physics (Thesis abstract) 31(10), 2935 (2004)

145. Lessard, E., Hsu, I.-C.J., Pouliot, J.: Inverse planning for interstitial gynecological template brachytherapy: truly anatomy based planning. Int. J. Radiat. Oncol. Biol. Phys. 54(5), 1243–1250 (2002)

146. Lessard, E., Pouliot, J.: Inverse planning anatomy-based dose optimization for HDR-brachytherapy of the prostate using fast simulated annealing algorithm and dedicated objective function. Med. Phys. 28(5), 773–779 (2001)

147. Lindemann, S.R., Hussein, I.I., LaValle, S.M.: Real time feedback control for nonholonomic mobile robots with obstacles. In: Proc. IEEE Conf. Decision and Control, pp. 2406–2411 (2006)

148. Liu, G.-R.: Mesh free methods: moving beyond the finite element method. CRC Press, Boca Raton (2003)

149. Lorensen, W.E., Cline, H.E.: Marching Cubes: A high resolution 3D surface construction algorithm. Computer Graphics (Proc. SIGGRAPH 1987) 21(4), 163–169 (1987)

150. Mamoudieh, A., Tremblay, C., Beaulieu, L., Lachance, B., Harel, F., Lessard, E., Pouliot, J., Vigneault, E.: Anatomy based inverse planning dose optimization in HDR prostate implant: A toxicity study. Radiotherapy and Oncology 75(3), 318–324 (2005)

151. Masamune, K., Ji, L., Suzuki, M., Dohi, T., Iseki, H., Takakura, K.: A newly developed stereotactic robot with detachable drive for neurosurgery. In: Wells, W.M., Colchester, A.C.F., Delp, S.L. (eds.) MICCAI 1998. LNCS, vol. 1496, pp. 215–222. Springer, Heidelberg (1998)

152. Mason, R., Burdick, J.: Trajectory planning using reachable-state density functions. In: Proc. IEEE Int. Conf. Robotics and Automation (ICRA), May 2002, vol. 1, pp. 273–280 (2002)

153. Maurin, B., Doignon, C., Gangloff, J., Bayle, B., de Mathelin, M., Piccin, O., Gangi, A.: CT-Bot: A stereotactic-guided robotic assistant for percutaneous procedures of the abdomen. In: Proc. SPIE Medical Imaging 2005, pp. 241–250 (2005)

154. Ménard, C., Susil, R., Choyke, P., Gustafson, G., Kammerer, W., Ning, H., Miller, R., Ullman, K., Crouse, N., Smith, S., Lessard, E., Pouliot, J., Wright, V., McVeigh, E., Coleman, C.N., Camphausen, K.: MRI-guided HDR prostate brachytherapy in a standard 1.5T scanner. Int. J. Radiat. Oncol. Biol. Phys. 59(5), 1414–1423 (2004)

155. Mendoza, C., Laugier, C.: Simulating soft tissue cutting using finite element models. In: Proc. IEEE Int. Conf. Robotics and Automation (ICRA), September 2003, pp. 1109–1114 (2003)

156. Moll, M., Goldberg, K., Erdmann, M., Fearing, R.: Orienting micro-scale parts with squeeze and roll primitives. In: Proc. IEEE Int. Conf. Robotics and Automation (ICRA), vol. 2, pp. 1931–1936 (May 2002)

157. Moore, A.W., Atkeson, C.G.: Prioritized sweeping: Reinforcement learning with less data and less real time. Machine Learning 13(1), 103–130 (1993)

158. Moore, A.W., Atkeson, C.G.: The parti-game algorithm for variable resolution reinforcement learning in multidimensional state spaces. Machine Learning 21(3), 199–233 (1995)

159. Morin, O., Gillis, A., Chen, J., Aubin, M., Bucci, M.K., Roach, M., Pouliot, J.: Megavoltage cone-beam CT: System description and IGRT clinical applications. Medical Dosimetry: Special Issue on Image-Guided Radiation Therapy (IGRT) 31(1), 51–61 (2006)

160. Morton, G.C.: The emerging role of high-dose-rate brachytherapy for prostate cancer. Clinical Oncology 17(4), 219–227 (2005)

161. Nash, S.G., Sofer, A.: Linear and Nonlinear Programming. McGraw-Hill, New York (1996)

162. Nath, R., Anderson, L.L., Luxton, G., Weaver, K.A., Williamson, J.F., Meigooni, A.S.: Dosimetry of interstitial brachytherapy sources: recommendations of the AAPM Radiation Therapy Committee Task Group No. 43. Med. Phys. 22(2), 209–234 (1995)

163. National Center for Health Statistics. Prostate disease (May 2004), http://www.cdc.gov/nchs/fastats/prostate.htm

164. Nienhuys, H.-W.: Cutting in Deformable Objects. PhD thesis, Utrecht University (June 2003)

165. Nienhuys, H.-W., van der Stappen, A.: A computational technique for interactive needle insertions in 3d nonlinear material. In: Proc. IEEE Int. Conf. Robotics and Automation (ICRA), April 2004, vol. 2, pp. 2061–2067 (2004)

166. Nienhuys, H.-W., van der Stappen, A.: A Delaunay approach to interactive cutting in triangulated surfaces. In: Boissonnat, J.-D., Burdick, J., Goldberg, K., Hutchinson, S. (eds.) Algorithmic Foundations of Robotics V (WAFR 2002), Springer Tracts in Advanced Robotics, pp. 113–129. Springer, Heidelberg (2004)

167. O'Brien, J.F., Hodgins, J.K.: Graphical modeling and animation of brittle fracture. Computer Graphics (Proc. SIGGRAPH 1999), 137–146 (August 1999)

168. Okamura, A.M., Simone, C., O'Leary, M.D.: Force modeling for needle insertion into soft tissue. IEEE Trans. Biomedical Engineering 51(10), 1707–1716 (2004)

169. Okazawa, S., Ebrahimi, R., Chuang, J., Salcudean, S.E., Rohling, R.: Hand-held steerable needle device. IEEE/ASME Trans. Mechatronics 10(3), 285–296 (2005)

170. O'Leary, M.D., Simone, C., Washio, T., Yoshinaka, K., Okamura, A.M.: Robotic needle insertion: Effects of friction and needle geometry. In: Proc. IEEE Int. Conf. Robotics and Automation (ICRA), September 2003, pp. 1774–1780 (2003)

171. Park, W., Kim, J.S., Zhou, Y., Cowan, N.J., Okamura, A.M., Chirikjian, G.S.: Diffusion-based motion planning for a nonholonomic flexible needle model. In. Proc. IEEE Int. Conf. Robotics and Automation (ICRA), April 2005, pp. 4611–4616 (2005)

172. Phee, L., Xiao, D., Yuen, J., Chan, C.F., Ho, H., Thng, C.H., Cheng, C., Ng, W.S.: Ultrasound guided robotic system for transperineal biopsy of the prostate. In: Proc. IEEE Int. Conf. Robotics and Automation (ICRA), April 2005, pp. 1327–1332 (2005)

173. Picinbono, G., Delingette, H., Ayache, N.: Nonlinear and anisotropic elastic soft tissue models for medical simulation. In: Proc. IEEE Int. Conf. Robotics and Automation (ICRA), May 2001, pp. 1370–1375 (2001)

174. Pickett, B., Vigneault, E., Kurhanewicz, J., Verhey, L., Roach, M.: Static field intensity modulation to treat a dominant intraprostatic lesion to 90 Gy compared to seven field 3-dimensional radiotherapy. Int. J. Radiat. Oncol. Biol. Phys. 44(4), 921–929 (1999)

175. Pouliot, J., Bani-Hashemi, A., Chen, J., Svatos, M., Ghelmansarai, F., Mitschke, M., Aubin, M., Xia, P., Morin, O., Bucci, K., Roach, I.M., Hernandez, P., Zheng, Z., Hristov, D., Verhey, L.: Low-dose megavoltage cone-beam CT for radiation therapy. Int. J. Radiat. Oncol. Biol. Phys. 61(2), 552–560 (2005)

176. Pouliot, J., Kim, Y., Lessard, E., Hsu, I.-C.J., Vigneron, D.B., Kurhanewicz, J.: Inverse planning for HDR prostate brachytherapy used to boost dominant intraprostatic lesions defined by magnetic resonance spectroscopy imaging. Int. J. Radiat. Oncol. Biol. Phys. 59(4), 1196–1207 (2004)

177. Pouliot, J., Lessard, E., Hsu, I.-C.J.: Advanced 3D Planning, 2nd edn., ch. 21. Medical Physics Publishing, Madison (2005)

178. Pouliot, J., Taschereau, R., Coté, C., Roy, J., Tremblay, D.: Dosimetric aspects of permanent radioactive implants for the treatment of prostate cancer. Physics in Canada 55(2), 61–68 (1999)

179. Renner, W.D., O'Conner, T.P., Bermudez, N.M.: An algorithm for generation of implant plans for high-dose-rate irradiators. Med. Phys. 17(1), 35–40 (1990)

180. Rivard, M.J., Coursey, B.M., DeWerd, L.A., Hanson, W.F., Huq, M.S., Ibbott, G.S., Mitch, M.G., Nath, R., Williamson, J.F.: Update of AAPM task group no. 43 report: A revised AAPM protocol for brachytherapy dose calculations. Med. Phys. 31(3), 633–674 (2004)

181. Roberson, P.L., Narayana, V., McShan, D.L., Winfield, R.J., McLaughlin, P.W.: Source placement error for permanent implant of the prostate. Med. Phys. 24(2), 251–257 (1997)

182. Satava, R.M.: Emerging technologies for surgery in the 21st century. Archives of Surgery 134(11), 1197–1202 (1999)

183. Satava, R.M.: Disruptive visions: predictive simulation–between scientific method and clinical trial is the role of modeling and simulation in scientific discovery and validation. Surgical Endoscopy 18(9), 1297–1298 (2004)

184. Satava, R.M.: Identification and reduction of surgical error using simulation. Minimally Invasive Therapy & Allied Technologies 14(4-5), 257–261 (2005)

185. Scheidler, J., Hricak, H., Vigneron, D.B., Yu, K.K., Sokolov, D.L., Huang, L.R., Zaloudek, C.J., Nelson, S.J., Carroll, P.R., Kurhanewicz, J.: Prostate cancer: Localization with three-dimensional proton MR spectroscopic imaging-clinicopathologic study. Radiology 213(2), 473–480 (1999)

186. Schneider, C., Okamura, A.M., Fichtinger, G.: A robotic system for transrectal needle insertion into the prostate with integrated ultrasound. In: Proc. IEEE Int. Conf. Robotics and Automation (ICRA), May 2004, pp. 2085–2091 (2004)

187. Schöberl, J.: Netgen - automatic mesh generator (2004), http://www.hpfem.jku.at/netgen/

188. Sellen, J.: Approximation and decision algorithms for curvature-constrained path planning: A state-space approach. In: Agarwal, P.K., Kavraki, L.E., Mason, M.T. (eds.) Robotics: The Algorithmic Perspective: 1998 WAFR, Natick, MA, pp. 59–67. AK Peters, Ltd. (1998)

189. Seymour, N.E., Gallagher, A.G., Roman, S.A., O'Brien, M.K., Bansal, V.K., Andersen, D.K., Satava, R.M.: Virtual reality training improves operating room performance: results of a randomized, double-blinded study. Annals of Surgery 236(4), 458–463 (2002)

190. Shechter, S., Schaefer, A.J., Braithwaite, S., Roberts, M., Bailey, M.: The optimal time to initiate HIV therapy. In: Annual Meeting of INFORMS (November 2005)

191. Shewchuk, J.: Triangle: A two-dimensional quality mesh generator and delaunay triangulator (2003), http://www-2.cs.cmu.edu/~quake/triangle.html

192. Shi, M., Liu, H., Tao, G.: A stereo-fluoroscopic image-guided robotic biopsy scheme. IEEE Trans. Control Systems Technology 10(3), 309–317 (2002)

193. Si, H.: Tetgen: A quality tetrahedral mesh generator and three-dimensional delaunay triangulator (Janurary 2006), http://tetgen.berlios.de/

194. Simone, C., Okamura, A.M.: Modeling of needle insertion forces for robot-assisted percutaneous therapy. In: Proc. IEEE Int. Conf. Robotics and Automation (ICRA), May 2002, pp. 2085–2091 (2002)

195. Tangelder, H., Fabri, A.: dD spatial searching. In: CGAL Editorial Board(ed.) CGAL-3.2 User and Reference Manual (2006)

196. Taschereau, R., Pouliot, J., Roy, J., Tremblay, D.: Seed misplacement and stabilizing needles in transperineal permanent prostate implants. Radiotherapy and Oncology 55(1), 59–63 (2000)

197. Taschereau, R., Stauffer, P., Hsu, I., Schlorff, J., Milligan, A., Pouliot, J.: Radiation dosimetry of a conformal heat-brachytherapy applicator. Technology in Cancer Research and Treatment 3(4), 347–358 (2004)

198. Taylor, R.G.: Models of Computation and Formal Languages. Oxford University Press, New York (1998)

199. Taylor, R.H., Stoianovici, D.: Medical robotics in computer-integrated surgery. IEEE Trans. Robotics and Automation 19(5), 765–781 (2003)

200. Terzopoulos, D., Platt, J., Barr, A., Fleischer, K.: Elastically deformable models. Computer Graphics (Proc. SIGGRAPH 1987) 21(4), 205–214 (1987)

201. The National Library of Medicine. Visible human project (2003),
     http://www.nlm.nih.gov/research/visible/visible%5Fhuman.html

202. Thrun, S., Burgard, W., Fox, D.: Probabilistic Robotics. MIT Press, Cambridge (2005)

203. Tubiana, M., Eschwege, F.: Conformal radiotherapy and intensity-modulated radiotherapy. Acta Oncologica 39(5), 555–567 (2000)

204. van den Berg, J., Overmars, M.: Planning the shortest safe path amidst unpredictably moving obstacles. In: Proc. Int. Workshop on the Algorithmic Foundations of Robotics (July 2006)

205. van Laarhoven, P.J.M., Aarts, E.H.L.: Simulated Annealing: Theory and Applications (Mathematics and Its Applications), 1st edn. D. Reidel Publishing Company (1987)

206. Vasquez, D., Large, F., Fraichard, T., Laugier, C.: High-speed autonomous navigation with motion prediction for unknown moving obstacles. In: Proc. IEEE/RSJ Int. Conf. on Intelligent Robots and Systems (IROS), pp. 82–87 (2004)

207. Waters, K.: A muscle model for animating three-dimensional facial expression. Computer Graphics (Proc. SIGGRAPH 1987) 21(4), 17–24 (1987)

208. Webster III, R.J., Kim, J.S., Cowan, N.J., Chirikjian, G.S., Okamura, A.M.: Nonholonomic modeling of needle steering. Int. J. Robotics Research 25(5-6), 509–525 (2006)

209. Webster III, R.J., Memisevic, J., Okamura, A.M.: Design considerations for robotic needle steering. In: Proc. IEEE Int. Conf. Robotics and Automation (ICRA), April 2005, pp. 3599–3605 (2005)

210. Webster III, R.J., Okamura, A.M., Cowan, N.J.: Toward active cannulas: Miniature snake-like surgical robots. In: Proc. IEEE/RSJ Int. Conf. on Intelligent Robots and Systems (IROS), pp. 2857–2863 (2006)

211. Webster III, R.J., Okamura, A.M., Cowan, N.J., Chirikjian, G.S., Goldberg, K., Alterovitz, R.: Distal bevel-tip needle control device and algorithm. US patent application number 11/436, 995 (May 2005)

212. Weissleder, R., Mahmood, U.: Molecular imaging. Radiology 219(2), 316–333 (2001)

213. Westergaard, H.M.: Theory of Elasticity and Plasticity. Dover Publications, Inc. (1964)

214. Wood, W.L.: Some transient and coupled problems - a state-of-the art review. In: Lewis, R.W., et al. (eds.) Numerical methods in transient and coupled problems, ch. 8. Wiley, New York (1987)

215. Wu, X., Dibiase, S.J., Gullapalli, R., Yu, C.X.: Deformable image registration for the use of magnetic resonance spectroscopy in prostate treatment planning. Int. J. Radiat. Oncol. Biol. Phys. 58(5), 1577–1583 (2004)

216. Wu, X., Downes, M., Goktekin, T., Tendick, F.: Adaptive nonlinear finite elements for deformable body simulation using dynamic progressive meshes. Eurographics 20(3), 349–358 (2001)

217. Xia, P., Pickett, B., Vigneault, E., Verhey, L., Roach, M.: Forward or inversely planned segmental multileaf collimator IMRT and sequential tomotherapy to treat multiple dominant intraprostatic lesions of prostate cancer to 90 Gy. Int. J. Radiat. Oncol. Biol. Phys. 51(1), 244–254 (2001)

218. Yan, D., Jaffray, D.A., Wong, J.W.: A model to accumulate fractionated dose in a deforming organ. Int. J. Radiat. Oncol. Biol. Phys. 44(3), 665–675 (1999)

219. Yershova, A., LaValle, S.M.: Improving motion-planning algorithms by efficient nearest-neighbor searching. IEEE Trans. Robotics 23(1), 151–157 (2007)

220. Yoo, T.S. (ed.): Insight into Images: Principles and Practice for Segmentation, Registration, and Image Analysis, 1st edn. AK Peters (2004)

221. Yu, Y., Zhang, J., Cheng, G., Schell, M., Okunieff, P.: Multi-objective optimization in radiotherapy: application to stereotactic radiosurgery and prostate brachytherapy. Artificial Intelligence in Medicine 19, 39–51 (2000)

222. Zaider, M., Zelefsky, M.J., Lee, E.K., Zakian, K.L., Amols, H.I., Dyke, J., Cohen, G., Hu, Y., Endi, A.K., Chui, C., Koutcher, J.A.: Treatment planning for prostate implants using magnetic-resonance spectroscopy imaging. Int. J. Radiat. Oncol. Biol. Phys. 47(4), 1085–1096 (2000)

223. Zhou, Y., Chirikjian, G.S.: Probabilistic models of dead-reckoning error in nonholonomic mobile robots. In: Proc. IEEE Int. Conf. Robotics and Automation (ICRA), September 2003, pp. 1594–1599 (2003)

224. Zhou, Y., Chirikjian, G.S.: Planning for noise-induced trajectory bias in nonholonomic robots with uncertainty. In: Proc. IEEE Int. Conf. Robotics and Automation (ICRA), April 2004, pp. 4596–4601 (2004)

225. Zhuang, Y.: Real-time simulation of physically realistic global deformations. PhD thesis, University of California, Berkeley (2000)

226. Zhuang, Y., Canny, J.: Real-time simulation of physically realistic global deformation. In: IEEE Vis 1999 Late Breaking Hot Topics (1999)

227. Zienkiewicz, O.C., Taylor, R.: The Finite Element Method, 5th edn. Butterworth-Heinemann (2000)

A Target Localization Using Deformable Image Registration

Recent advances in medical imaging are enabling physicians to non-invasively pinpoint the location of cancerous cells inside the body. But after obtaining diagnostic images in which the cancer is localized, the patient is generally treated in a different facility, days, weeks, or months later. During this time, the patient may experience substantial changes, such as tumor size changes or weight changes that affect the location of the cancer cells relative to markers on the patient's skin. Furthermore, due to the clinical constraints of the diagnostic imaging modality and treatment procedure, the patient's position may be different between the diagnostic and and treatment phases. For radiation cancer treatment, these patient changes due to time, movement, and imaging modality constraints can lead to misalignment of the radiation dose, reducing the conformality of dose to the tumor and resulting in suboptimal treatment [11, 29, 96].

In this appendix, we develop an image registration approach that explicitly considers tissue deformations and variations in model parameters between patients to improve target localization across images acquired at different times. The method, which is based on a biomechanical model of soft tissue deformation, combines a nonlinear optimization algorithm with results from physically-based soft tissue simulation described in chapter 2.

We apply the new method to register diagnostic MRSI prostate images with radiation treatment planning images. To obtain a sufficient signal-to-noise ratio for MRSI, a probe must be placed near the prostate, which results in substantial deformations of the surrounding soft tissues, as shown in figure A.1. This probe must be removed during treatment. Results for 10 prostate cancer patient cases indicate that our method provides a statistically significant improvement in target registration accuracy compared to past methods [11].

A.1 Introduction to Deformable Image Registration

Registration is the process of finding a spatial transform that maps points from one image to the corresponding points in another image of the same subject [220]. The input data to the image registration process is two images: the first image

R. Alterovitz and K. Goldberg: Motion Planning in Medicine, STAR 50, pp. 129–147, 2008.
springerlink.com

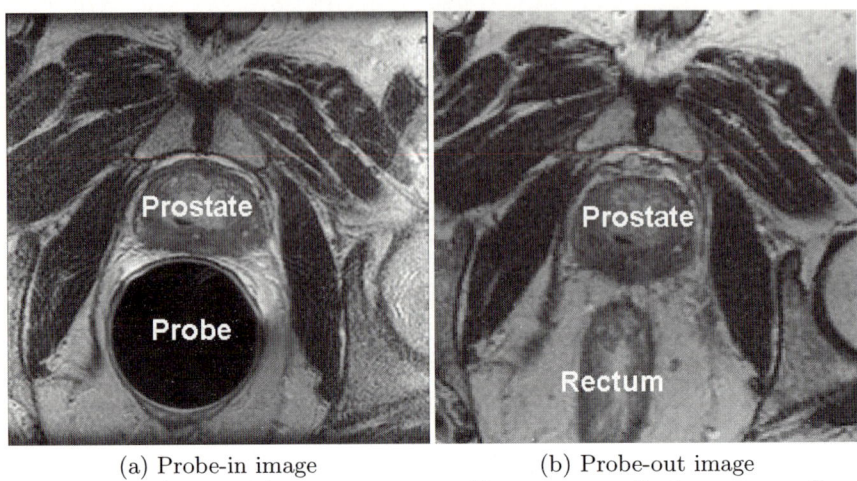

| (a) Probe-in image | (b) Probe-out image |
| (for MRSI) | (for prostate radiation treatment) |

Fig. A.1. MRSI data for the prostate is obtained with a balloon endorectal probe, as shown in an axial MR image at the mid-gland of the prostate (a). Radiation treatment is performed with the probe removed (b). The balloon endorectal probe causes substantial deformation of the prostate.

is defined as the *fixed image* (or *reference image*) F and the second image is defined as the *moving image* (or *deforming image*) M. The goal of registration is to determine a spatial transform T that will align the moving image with the fixed image.

Most image registration methods, including the method developed in this appendix, use a software framework consisting of a transform, a metric, and an optimizer [220]. The transform T, parameterized by a set of transform parameters $\mathbf{p}$, defines a mapping of points from the fixed image onto the moving image. This transform can be used to generate a transformed moving image M' by applying the the transform $\mathbf{x}' = T(\mathbf{x}|\mathbf{p})$ for each pixel coordinate $\mathbf{x} \in F$ and setting the pixel intensity at pixel coordinate $\mathbf{x}$ of M' to the pixel intensity at $\mathbf{x}'$ of M. The metric $S(\mathbf{p}|F, M, T)$ measures the similarity of the the transformed moving image M' with the fixed image F. An optimizer searches over the space of all feasible transform parameters $\mathbf{p}$ to maximize the quantitative registration quality criterion defined by the metric S so M' matches F as best as possible.

In *rigid registration*, we assume that the spatial transform T that maps points from the fixed image to the moving image is a rigid body transform: it includes only translation and rotation, as shown in figure A.2. Given a point $\mathbf{x}$ in F, the corresponding point in M is given by $\mathbf{x}' = T(\mathbf{x}|\mathbf{p})$. In 2-D, the parameter vector $\mathbf{p}$ for a rigid transform $T(\mathbf{p})$ has a dimension of 3 (x-axis translation, y-axis translation, and rotation by θ degrees in the xy-plane). When registering rigid 3-D volumes, $\mathbf{p}$ has a dimension of 6 (3 translation degrees of freedom and 3 rotation degrees of freedom) [220].

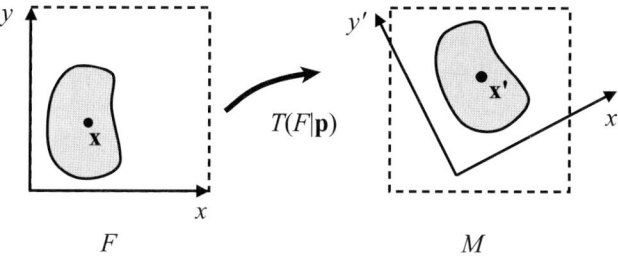

Fig. A.2. Fixed image F and a moving image M are enclosed in dashed lines. The object of interest in F is translated and rotated in M. The transformation T defines a rigid transformation that maps points $\mathbf{x}$ from F to $\mathbf{x}'$ in M.

However, soft tissues may deform between image acquisitions due to causes such as patient position changes, physiological changes such as bladder volume changes, and imaging requirements such as probes. In these cases, the assumption of a rigid transform is no longer valid. In *non-rigid* or *deformable registration*, we do not assume that the transform T is limited to translation and rotation. Instead, T can represent an arbitrary mapping from F to M, as shown in figure A.3.

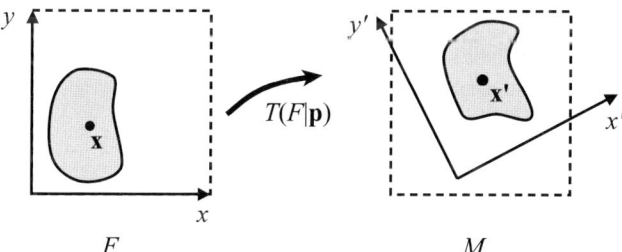

Fig. A.3. Fixed image F and a moving image M are enclosed in dashed lines. The object of interest in F is translated, rotated, and deformed in M. The transformation T defines a deformable transformation that maps points $\mathbf{x}$ from F to $\mathbf{x}'$ in M.

In the most general case, a deformable transform T would define a mapping individually for every pixel in F to its corresponding location in M. For 3-D images with x, y, z pixel dimensions of d_x, d_y, d_z, respectively, this would result in a parameter set $\mathbf{p}$ of dimension $3 d_x d_y d_z$. Since medical images today are typically of the dimensions $256 \times 256 \times 256$, optimizing a parameter vector of dimension $3 \cdot 256^3 = 50,331,648$ variables with a possibly nonlinear, non-convex objective metric S is computationally intractable.

To limit the dimension of $\mathbf{p}$ and make the deformable registration problem computationally tractable, numerous methods have been developed that explicitly compute transformations for a subset of pixels in the image, and then intelligently interpolate the transformation for the remaining pixels in the image. Methods based on this principle include B-splines [220], energy models [215], viscous models [53], and elastic models [26]. Each of these methods implicitly makes assumptions about the types of deformations that occur in the subject of the image being registered.

In this appendix, we develop a deformable registration algorithm specifically designed for medical images that explicitly considers soft tissue deformations and variations in model parameters between patients. Rather than relying on a mathematical abstraction such as B-splines that has no physical basis, our approach is based on building a biomechanical model of the tissue in the image. By explicitly modeling the underlying anatomical structures, our approach is physically based.

The core of our image registration method is a biomechanical model based on a finite element method. In 1982, Bajcsy and Broit performed pioneering work in the application of elastic finite element models to deformable image registration [26]. Initial methods used a regular finite element grid and computed a set of external forces that deform the grid to minimize a function defined by an elastic energy and a similarity energy. This work was later extended to include a rigid registration pre-processing step and multi-resolution hierarchical registration [220]. Advances in geometric algorithms and computation speed are now enabling the generation of finite element meshes that conform to tissue type boundaries and simulations that explicitly model sources of large soft tissue deformations in just seconds or minutes of computation time. Recent work has modeled large deformations such as the compression of breast tissue for biopsies [25], as well as our work on prostate deformations due to forces exerted by needles during insertion into soft tissue [18].

However, biomechanical models require knowledge of tissue material properties such as stiffness and compressibility. In past work on deformable image registration that we are aware of, tissue material properties are either fixed as constants for all patients [35, 60, 218] or implicitly held constant across an entire image [215]. After building a biomechanical model, we include uncertain parameters such as tissue material properties in $\mathbf{p}$ as variables. This allows our optimizer to estimate these uncertain parameters and maximize deformable image registration quality. When applied to prostate MRS/MR images, our new method results in a significant improvement over previous methods, as discussed in section A.3.5.

A.2   Deformable Registration with Model Parameter Estimation

Our image registration method defines a transform T that maps points between a fixed image and a moving image. Given a fixed image F and a moving image M, the goal is to compute parameters $\mathbf{p}^*$ such that $T(F|\mathbf{p}^*) = M$. In our method, T is invertible. The inverse mapping T^{-1} transforms every point in the moving image M to its coordinate in the fixed image F.

At the core of our method to compute T is a finite element method that estimates the deformation of soft tissues in the fixed image due to known external forces or constraints. Treating the uncertain tissue stiffness properties and external forces as unknown variables, we estimate their values using nonlinear local optimization to maximize image registration quality.

A.2.1   Method Input

The input for our image registration method includes a fixed image F and moving image M. To build a biomechanical model, the method also requires polygonal segmentation boundaries of distinct tissue types in the images. This information is typically already available in radiation oncology applications since segmentation is required for dose planning. Although segmentation is usually performed by hand for reliability in clinical practice, methods are being developed to automatically segment tissue types [220]. Each segmented region in the images must be labeled with a corresponding tissue type, such as bone or muscle. We let m be the number of distinct tissue types in the fixed image F.

Our method also optionally accepts as input known constraints on tissue deformation between the images. These constraints are specified as a set H of homologous point pairs. For each pair $(\mathbf{x}, \mathbf{x}') \in H$, $\mathbf{x}$ is a point in the fixed image and $\mathbf{x}'$ is the coordinate of the homologous point in the moving image. For images in which no bones are present, we require $|H| \geq 1$ to ensure that the linear system of equations defined by the finite element method in section A.2.2 is solvable.

Our method also optionally accepts as input a set L of points or polygonal tissue type boundaries that may be subject to external forces of unknown magnitudes. For example, points on the segmented boundary of the bladder should be listed since the bladder may expand or contract between image acquisitions.

A.2.2   Finite Element Model of Soft Tissue Deformation

As discussed in chapter 2, we approximate soft tissues as nearly incompressible (Poisson's ratio of 0.49), linearly elastic, and isotropic. Although tissue stiffness properties and external forces will be modified during the optimization method, initial default values must be set in the transform parameter vector $\mathbf{p}$. Based on tissue stiffness measurements obtained using ultrasound elastography [126], we temporarily assign a Young's modulus of 30 kPa to all soft tissues and assume bones are rigid. We assume initial external forces all have zero magnitude.

We automatically generate a finite element mesh that conforms to the segmented tissue boundaries for the fixed image F. For 2-D, we generate triangular elements using the constrained Delaunay triangulation software program *Triangle* [191]. For 3-D, we use the tetrahedral mesh generation software *TetGen* [193]. elements in the mesh are assigned default stiffness properties. Mesh nodes defining elements inside bones are constrained to be fixed. We let l be the number of nodes along boundaries included in set L.

We use the finite element method (FEM) to estimate tissue deformations. The homologous points $(\mathbf{x}, \mathbf{x}') \in H$ specify displacement constraints for the finite element method, where the node at point $\mathbf{x}$ is displaced by $\mathbf{x}' - \mathbf{x}$ and constrained as fixed. The deformations of the surrounding soft tissues are then computed using FEM. As described in chapter 2, the FEM problem for a given N-D fixed image mesh with n nodes is defined by a system $\mathbf{Ku} = \mathbf{f}$ containing Nn linear equations where $\mathbf{K}$ is the global stiffness matrix, $\mathbf{f}$ is the external force vector, and $\mathbf{u}$ is the nodal displacement vector. For each fixed node, we remove its N corresponding degrees of freedom from the system. We solve the linear system of equations numerically using the Gauss-Seidel method to compute nodal displacements $\mathbf{u}$ for non-fixed nodes. By using linear interpolation within each element of the mesh, the nodal displacement vector $\mathbf{u}$ defines a complete invertible mapping function T between the fixed image and the moving image. The mapping T is applied to every pixel in the fixed image F to obtain the deformed fixed image $T(F|\mathbf{p})$.

A.2.3   Quality Metric

We use a quality metric S to quantify how closely the deformed fixed image $T(F|\mathbf{p})$ matches the moving image M. Any image similarity metric S can be used, including image intensity metrics such as mutual information, homologous point distance measures, and segmented region overlap metrics [220]. However, the computation time and convergence properties of the optimization algorithm defined in section A.2.4 depend on the quality metric.

A.2.4   Optimization of Uncertain Parameters

Our image registration method treats tissue stiffness properties and external forces at user-specified nodes as uncertain parameters. The stiffness for soft tissue is constrained between Y_{min} and Y_{max}, where we select $Y_{min} = 1$ kPa and $Y_{max} = 600$ kPa as limits based on tissue elastography results [126]. External force magnitudes are unbounded. We define the optimization objective function for maximization as:

$$Q = S - \alpha E$$

where S is the selected quality metric, α is a scaling parameter, and E is the percent of strain energy due to external forces. To compute E, the optimization algorithm computes tissue deformations twice, first without external forces and then with external forces added. For each case, it computes the total strain by summing the strain of each element in the mesh, which is quickly computed by multiplication of element stiffness matrices and vectors of node displacements [227]. We subtract αE in the objective function to prioritize optimization of parameters of the physically-based model (tissue stiffness) relative to external forces. Appropriately setting α, which is problem specific, produces visually smoother image mappings by preventing unrealistic large magnitude external forces.

We apply the Steepest Descent method with Armijo's Rule for line search [31] to maximize the nonlinear objective function Q. The variables, which include m tissue stiffness properties and l external force degrees of freedom, are defined in a vector $\mathbf{p}$ of dimension $m + l$. The quality metric Q is a function $Q(\mathbf{p})$. We numerically compute derivatives for the gradient $\nabla Q(\mathbf{p})$ using finite differences with sufficiently high differences to avoid numerical difficulties. At iteration i of the Steepest Descent optimization method, Armijo's Rule selects the next candidate point $\mathbf{p}_{i+1} = \mathbf{p}_i + 2^t \lambda \nabla Q(\mathbf{p}_i)$ for predefined step size λ by sequentially incrementing integer t starting at $t = 0$ to solve for the maximum t that improves $Q(\mathbf{p}_{i+1})$. Then the gradient $\nabla Q(\mathbf{p}_{i+1})$ is computed and the Steepest Descent algorithm repeats until iteration j where $\|\nabla Q(\mathbf{p}_j)\| < \epsilon$ for $\epsilon = 0.001$. Because the objective function Q is not guaranteed to be convex, this method may not find a global optimal solution [31]. We label the local optimal solution found as $\mathbf{p}^*$.

A.2.5   Visualizing Registration Output

In 2-D, rendering the deformed fixed image can be performed quickly using texture-mapping, as described in chapter 2. We take advantage of the fact that the interpolation function inside triangular elements for linearly elastic finite element methods is linear, as described in chapter 2. Rather than explicitly applying the mapping T to every pixel in the fixed image F, we instead only compute the transformation $T(\mathbf{x_i}|\mathbf{p})$ for nodes $\mathbf{x_i}$ in the mesh, and use hardware accelerated texture-mapping to linearly interpolate pixel transformations for pixels inside the mesh elements. We are currently developing computationally efficient visualization methods for 3-D deformable image registration results.

A.3   Application to Prostate Cancer Treatment

We apply our deformable registration method to prostate cancer treatment. In 1996, Kurhanewicz et al. showed that magnetic resonance spectroscopic imaging (MRSI), a type of functional imaging that measures concentration of metabolic compounds, can be used to noninvasively diagnose and locate cancerous tumors in the prostate [129, 130, 131, 185]. By measuring choline, polyamine, and citrate levels which change with the evolution and progression of cancer, MRSI can be used to identify the location and extent of dominant intraprostatic lesions (DIL's) in the prostate [128]. Combining magnetic resonance imaging (MRI) with MRSI allows identification of a tumor with specificity of up to 91% [185].

Knowledge of cancer location can assist physicians during radiation treatment planning. Numerous studies indicate that improving the conformality of radiation dose to the cancer location significantly improves cancer treatment and reduces negative treatment side effects [110, 127, 203]. Physicians can escalate the radiation dose to the cancer location using treatment methods such as HDR brachytherapy [176], permanent seed brachytherapy [222], and external beam radiation treatment [174, 217].

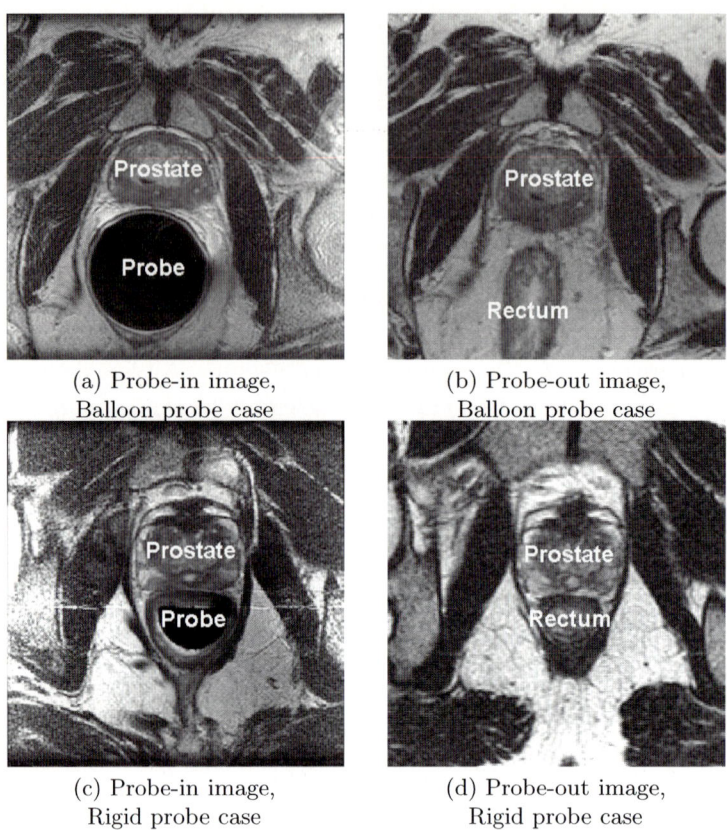

(a) Probe-in image,
Balloon probe case

(b) Probe-out image,
Balloon probe case

(c) Probe-in image,
Rigid probe case

(d) Probe-out image,
Rigid probe case

Fig. A.4. MRSI data for the prostate is obtained with a balloon endorectal probe inserted and inflated (a) or a rigid endorectal probe (c) as shown in the axial MR images at the mid-gland of the prostate. Radiation treatment is performed with the probe removed (b), (d).

To obtain improved signal-to-noise ratio (SNR) and better spatial resolution MRI and MRSI, an endorectal probe integrated with a pelvic phased array (PPA) coil is commonly used. The endorectal probe is critical for the acquisition of high spatial resolution ($\approx$0.3 cc) MRSI data of the prostate due to the approximate 10-fold increase in SNR relative to external phased array coils [98, 104, 108, 129, 130, 185]. However, the probe may cause considerable nonlinear translation and distortion of the prostate [122], as shown in figure A.4. The probe is generally removed during imaging for radiation treatment planning and therapy. To effectively utilize the MRSI data, clinicians must register the probe-in image to a probe-out image.

To register probe-in images obtained during a combined MRI/MRSI staging examination to probe-out images, we apply our deformable registration method described in section A.2. We use a 2-D finite element model and estimate the

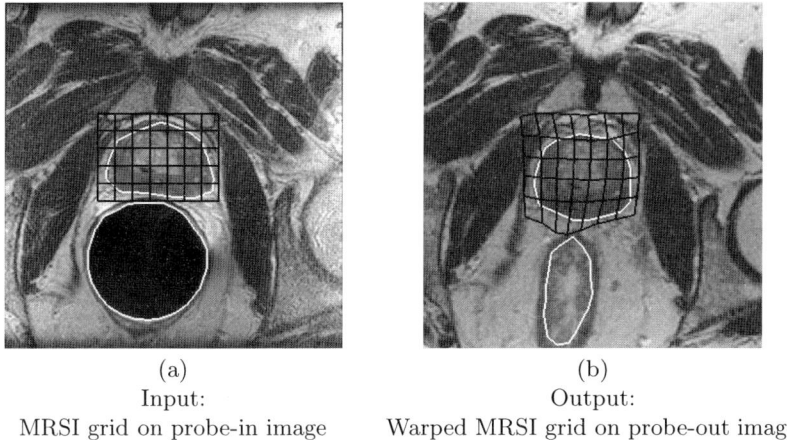

(a)

Input:

MRSI grid on probe-in image

(b)

Output:

Warped MRSI grid on probe-out image

Fig. A.5. Spectroscopy data is obtained for voxels inside the MRSI grid overlaid on an the axial probe-in MR image (a). Our image registration method warps the MRSI grid to the probe-out image for use during treatment planning (b).

deformation of the prostate and surrounding tissues in the plane of the image due to the insertion of an endorectal probe. A 2-D model is sufficient for our application since the out-of-plane deformations are smaller than the thickness of imaging slices [59, 122]. However, patient-specific model parameters required as input for the biomechanical model are not known with certainty, including tissue stiffness properties for the prostate and surrounding soft tissues. Additional uncertain parameters include forces due to patient position changes, bladder volume changes, and other factors that differ between the probe-out and probe-in images but are not explicitly included in our linear elasticity soft tissue deformation model. As described in section A.2, we use a local nonlinear optimization algorithm to estimate uncertain patient-specific tissue stiffness properties and external forces to maximize image registration quality. Compensating for computed tissue deformations results in a nonlinear warping of the MRSI grid, as shown in figure A.5.

Past work on image registration of the prostate includes rigid transformations [80, 122], spline transformations [79, 120], energy models [215], and finite element models [35, 60, 218] for registering dose calculation CT images [218], treatment and interventional MR images [35, 79, 80], probe-in/probe-out MR images [60], and MR images with endorectal balloons at different levels of inflation [215]. Fei et al. ignore tissue deformations that occur between pre-operative and interventional MR images and maximize the mutual information (MI) or correlation coefficient (CC) of the image intensity histograms using rigid body translation and rotation of the prostate [80]. Kim et al. rigidly align probe-in and probe-out images by first rotating the images into the same plane, then doing a rigid 2-D translation in plane [122]. For a sample probe-in/probe-out case in which prostate deformation is minimal, this method achieves less than 2 mm registration error. Fei et al. and Kessler et al. use spline methods, which non-linearly warp an image using a non-physically based model with a large number of degrees of freedom [79, 215]. They use multiresolution approaches to increase avoidance of local maxima of the CC and MI metrics. Wu et al. develop a hybrid method for registering MR images with endorectal balloons at different levels of inflation by maximizing an objective function containing a weighted sum of MI and regularization energy from a non-finite element physically based model [215]. A key advantage of these methods based on MI and CC quality metrics is that tissue segmentation is not required, but these methods have large numbers of degrees of freedom, are prone to local maxima, require long computation times (18-22 minutes for Wu et al.), and have potentially larger error due to soft boundaries of deformable tissues [80, 215]. MI and CC metrics cannot be applied in isolation to our problem of registering a probe-in image to a probe-out image because, without segmentation, the probe-out image contains no information on the probe insertion location. Physically based biomechanical models, such as the finite element method, have potential to address some of these limitations. Finite element methods require image segmentation to define tissue type boundaries (to specify tissue-specific material properties) and mesh generation. Yan et al. performed pioneering work in deformable image registration based on the finite element method to calculate fractionated dose in a deforming organ [218]. They segmented a single tissue type, the rectal wall, and applied the method to inter-treatment motion using fiducials to set boundary conditions. Bharatha et al. and Crouch et al. apply linear elasticity finite element modeling to the prostate using a tetrahedral mesh with distinct central gland and peripheral zone regions [35] and a hexahedral mesh using a medially-based solid representation with uniform tissue properties inside the prostate [60]. Image registration based on biomechanical models, including finite element and energy methods, require tissue material properties as input. In past work we are aware of, material properties are either fixed as constants for all patients [35, 60, 218] or implicitly held constant across an entire image [215]. Our image registration method uses nonlinear optimization to set patient-specific values for uncertain parameters in the biomechanical model including separate tissue stiffness values for each segmented tissue type [8, 9, 11, 12]. We also explicitly warp MRSI grids to compensate for tissue deformations.

A.3.1 Patient Image Acquisition

We applied our image registration method retrospectively to 10 patient cases. The patients were recruited from January to June, 2003, at the Magnetic Resonance Science Center (MRSC), University of California, San Francisco. A balloon probe (USA Instruments, Aurora, OH) with 100 cc of air injected was used for 5 patients while a rigid probe (MedRad, Pittsburgh, PA) was used for the remaining 5 patients. Once inflated, the balloon probe had a circular cross-section with a 48 mm diameter. The rigid probe was a half ellipse, in cross-section, with the anterior surface flat. Its right - left extent was 29 mm and its anterior - posterior extent was 16.5 mm. Combined MRI/MRSI was obtained using the balloon or rigid probe in combination with an external phased array of coils on a 1.5 Tesla GE system (Signa, GE Medical Systems, Milwaukee, WI). For rigid probe cases, the USA torso phased array was used, while the GE pelvic phased array was used for the balloon probe cases. These two probes were selected because the MedRad coil is the only MR probe currently available commercially and the USA Instruments probe is commercially manufactured and will soon be a commercially available alternative probe.

The probe-in images used in this study were acquired during a "PROSE" (PROstate Spectroscopy and imaging Examination) MRI/MRSI examination (GE Medical Systems, Milwaukee, WI). The details of the MR imaging method used have been discussed in previous work [90, 122, 130, 176]. Spectroscopy data was obtained for $7 \times 7 \times 7$ mm voxels ($\approx$0.3 cc). Thin-section high spatial resolution axial T_2 weighted fast spin-echo images of the prostate and seminal vesicles were obtained with a slice thickness of 3 mm, an inter-slice gap of 0 mm, and a field of view (FOV) of 14 cm. At the end of the "PROSE" MRI/MRSI examination, the endorectal probe was removed with the patient remaining on the imaging table. Additional sagittal and axial fast spin echo T_2 weighted images were acquired without the endorectal probe using the phased array coil alone for signal reception. As with the probe-in case, patients were scanned in the supine position. All image acquisition parameters for the probe-out images were the same as for the probe-in images except for increasing the field-of-view (FOV) from 14 cm to 20 cm to partially compensate for the reduction in SNR obtained without the use of an endorectal probe.

A.3.2   Application of the Deformable Registration Method

We apply the deformable registration method described in section A.2 to the prostate images described in section A.3.1 We define the fixed image F as the probe-out image and define the moving image M as the probe-in image. The transform T attempts to mimic the deformation of the prostate and surrounding tissue due to endorectal probe insertion. Given a probe-out image F and a probe-in image M, the goal is to compute parameters $\mathbf{p}$ such that $T(F|\mathbf{p}) = M$. The inverse mapping T^{-1}, which can be used during treatment planning, transforms every point in the MRSI grid of the probe-in image M to its coordinate in the probe-out image F.

From the probe-in and probe-out MR image volumes, we selected a single probe-in image slice M at the mid-gland of the prostate for each patient. We then manually selected a corresponding probe-out image slice F that is at the same level as the probe-in image for the patient. As a pre-processing step, we rigidly register the images by aligning points on non-deforming tissues, such as points in bones, using a homologous point method to translate the images [102].

We manually segmented the selected images using a standard image segmentation method by drawing polygonal outlines on a computer screen to define the boundaries of tissue types. For cases in which the tissue type (such as the rectum) was close to circular, we specified a circle and radius that the software automatically converted to a polygonal approximation. The image registration

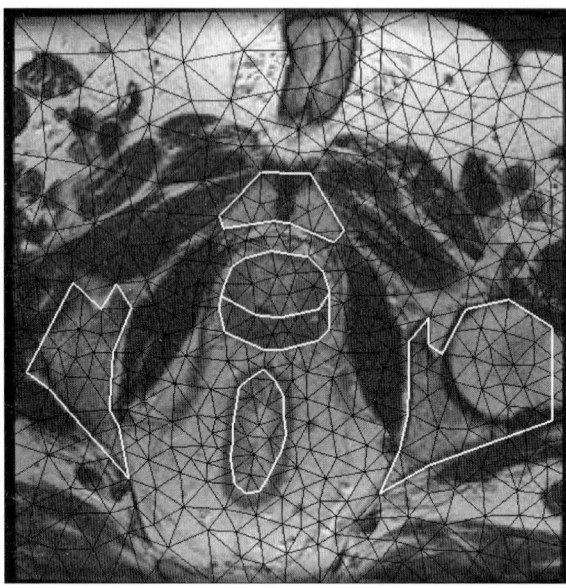

Fig. A.6. Conformal Delaunay triangular mesh (black triangles) for a probe-out image with central gland and peripheral zone of the prostate, probe entry location (rectum), and bones segmented (in white)

method requires segmentation of the probe and prostate in the probe-in image and the probe entry location (rectum) and prostate in the probe-out image. For improved accuracy in the biomechanical simulation, we also segmented bones and separately segmented the central gland (CG) and peripheral zone (PZ) of the prostate in the probe-out image. Additional segmentation of the probe-in image will not improve results since the biomechanical model is applied to deform the probe-out image.

For this application, the known constraints on deformation are the displacements caused by the endorectal probe. As shown in figure A.7, our model expands the rectum lining in the probe-out image to match the probe outline in the probe-in image. We project points along the probe outline in F along the ray based at the rectum center and constrain them to the intersection with the probe outline in M. This defines the set of homologous points H. We define set L as the prostate gland boundary.

We automatically generate a finite element mesh composed of $n = 500$ nodes and between 800 and 1,000 triangular elements using *Triangle* [191]. Image segmentation and the mesh generated for the probe-out image of a sample case are shown in figure A.6.

Based on tissue stiffness measurements obtained using ultrasound elastography in previous work [126], we assign a Young's modulus of 60 kPa to the central gland of the prostate and 30 kPa to all surrounding tissues for all patient images during initialization of the method.

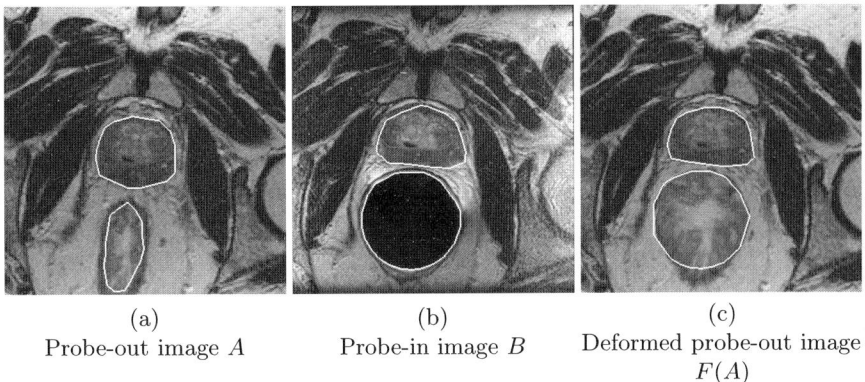

(a)	(b)	(c)
Probe-out image A	Probe-in image B	Deformed probe-out image $F(A)$

Fig. A.7. Probe-out image A with segmented prostate gland (outlined in white, middle) and rectum (outlined in white, bottom) (a) and the corresponding probe-in image B with prostate and probe segmented (b). The method computes image $F(A)$ (c) which displaces mesh nodes along the rectum in the probe-out image to the probe outline in the probe-in image and estimates the resulting soft tissue deformations. The image registration quality (DSC value) between (b) and (c) is 97.8%.

The number of distinctly segmented soft tissue types in F was $m = 3$. After meshing, the number of nodes in set L is typically between 20 and 40. This results in a parameter vector $\mathbf{p}$ of size between 43 and 83.

We define the metric S using the Dice Similarity Coefficient (DSC), a metric that measures overlap of polygonal regions. For this application, we measure the overlap between the prostate area in the probe-in image M and the prostate area in the deformed probe-out image $T(F|\mathbf{p})$ using the Dice Similarity Coefficient (DSC). Superimposing an outlined area from two images, the DSC is defined as:

$$D = \frac{2a}{2a + b + c}$$

where a is the number of picture elements (pixels) shared by both areas, b is the number of pixels unique to the first area, and c is the number of pixels unique to the second area [35, 65]. The DSC is a scalar between 0 and 1 with higher values representing better quality registration.

A.3.3   Warping the MRSI Grid

The 3-D MRSI data is collected from a volume and individual spectra are generally reconstructed for $7 \times 7 \times 7$ mm voxels within a grid overlaid on this volume. To help register spectroscopic data to the probe-out image, we transform each intersection point in the regular MRSI grid from the probe-in image plane to the probe-out image using the inverse of mapping T. The warped MRSI grid is the output of the algorithm: it registers the probe-in MRSI data to a probe-out image for use during treatment planning.

A.3.4   Method Evaluation and Parameter Selection

We evaluate the image registration method using two metrics: DSC and point error. We compute the DSC using the prostate outline in the probe-in image M and the prostate outline in the deformed probe-out image $T(F|\mathbf{p}^*)$. We compare our deformable image registration method to a rigid registration method where the center of mass of the prostate total gland is translated in the probe-out image by the distance between its center of mass in the probe-out and probe-in images [35].

As a second measure of image registration quality, we evaluate displacement errors of homologous points in the interior of the prostate on the probe-in images M and the deformed probe-out images $T(F|\mathbf{p}^*)$. As in past work by Bharatha et al. [35], we select points on the probe-in images at the posterior border of the central gland near the midline of the prostate. We then select homologous points corresponding to the same tissue location on the probe-out images using patient-specific local image pixel intensity variations as references. Our image registration method maps the point on the probe-out image F to the deformed probe-out image $T(F|\mathbf{p}^*)$ so we can directly measure the point error: the distance between the homologous point in M and $T(F|\mathbf{p}^*)$. We compare this error to the distance between the homologous points in the given probe-in image M and probe-out image F to quantify the registration improvement resulting from the method.

Two parameters of the method that influence image registration quality and must be set are n, the number of nodes in the mesh, and α, the scaling parameter in the objective function Q that weighs direct maximization of the DSC relative to the percent of strain energy E due to external forces. For a subset of the patient data (3 balloon probe cases and 3 rigid probe cases), we evaluated image registration quality for $n = 100, 500$, and $1,000$ and for $\alpha = 0.0, 0.005$, and 0.01.

A.3.5   Results

The mean DSC of our method was 97.5% with a standard deviation of 0.7% for the 5 balloon probe cases. For the 5 rigid probe cases, the mean DSC was 98.1% with a standard deviation of 0.4%. As shown for a patient case in figure A.7, the deformed probe-out image closely matches the probe-in image. In Table A.1, we compare our image registration method to rigid registration based on center-of-mass translation for the prostate total gland. We performed paired t-tests to determine the statistical significance ($P < 0.05$) of the results and found that the improvement in DSC using our method was statistically significant for both the balloon probe ($P = 0.035$) and the rigid probe ($P = 0.013$) cases.

The results of our method for the point error metric are shown in Table A.2. Our method reduces displacement error between the homologous points in the probe-in and probe-out images by a mean of 74.8% to a mean error of 1.95 mm for the balloon probe cases. For the rigid probe cases, the reduction was by a mean of 70.0% to a mean error of 0.97 mm. We performed paired t-tests and found that the reduction in error was statistically significant for both the balloon probe ($P = 0.0045$) and the rigid probe ($P = 0.0099$) cases.

Table A.1. DSC mean and standard deviation (in parentheses) for image registration quality

	Rigid Translation	Our Method
5 balloon probe cases	86.6% (10.4%)	97.5% (0.7%)
5 rigid probe cases	86.6% (10.4%)	97.5% (0.7%)

Table A.2. Point displacement error means and standard deviations (in parentheses) for sample homologous points on the boundary of the prostate central gland and peripheral zone near the midline

	Mean point error for probe-in / probe-out images (mm)	Mean point error after our method (mm)	Mean reduction in error (%)
5 balloon probe cases	9.22 (3.22)	1.95 (0.22)	74.8% (15.1%)
5 rigid probe cases	3.93 (1.59)	0.97 (0.51)	70.0% (27.2%)

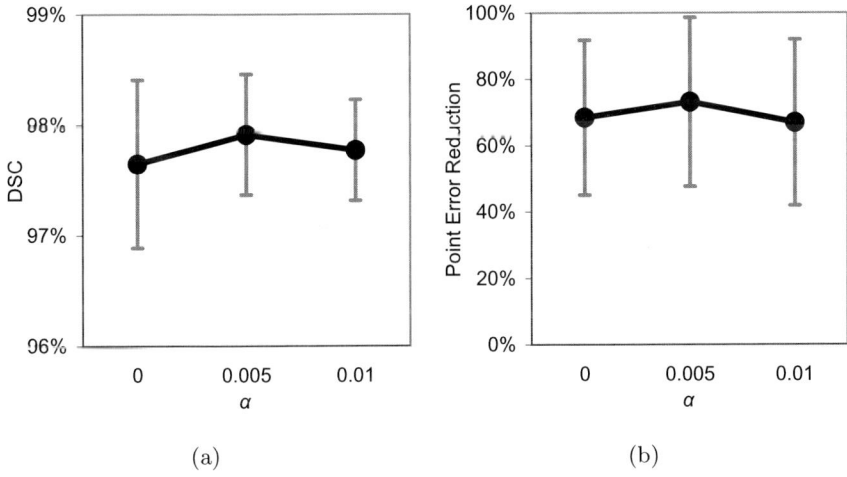

(a)   (b)

Fig. A.8. Sensitivity of mean image registration quality (DSC and point errors) to the optimization parameter α, with error bars for standard deviations

For these results, we set parameter α in the formula for objective function Q in section A.2.4 to 0.005. Decreasing α allows for greater external forces while increasing α penalizes external forces in favor of tissue stiffness during optimization of uncertain parameters. The trade-off effect of α on DSC and point error is shown in figure A.8(a) and (b). Increasing α to 0.01 or decreasing α to 0.0 results in lower mean DSC and higher mean point errors.

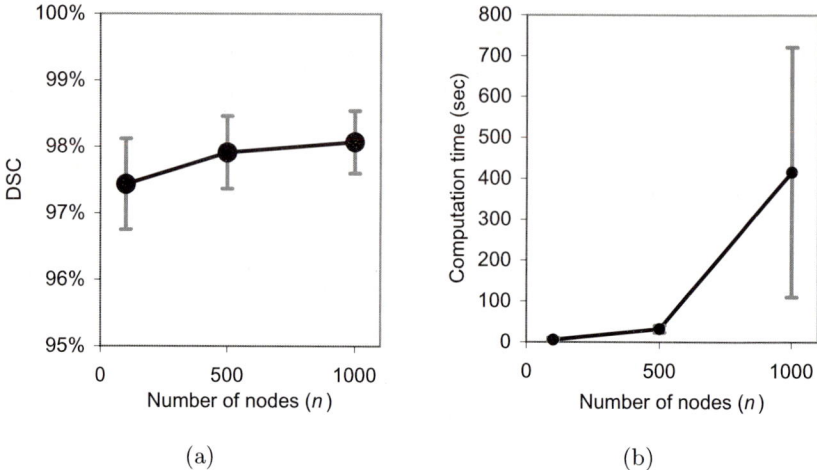

Fig. A.9. Sensitivity of mean DSC image registration quality (a) and computation time (b) to the number of nodes n in the mesh, with error bars for standard deviations

We also performed sensitivity analysis on n, the number of nodes in the finite element mesh. Increasing n improves average image registration quality measured by DSC, as shown in figure A.9(a). However, this improvement comes at a large computation cost, as shown in figure A.9(b), with 1,000 node meshes requiring over 6 minutes of computation time on average. Results in this study use meshes with $n = 500$ nodes, which requires less than 1 minute of computation time per image slice while maintaining good image registration quality; DSC results with $n = 500$ are not significantly different from DSC results with $n = 1,000$ ($P = 0.324$).

We show the output of our image registration method for a sample balloon probe patient in figure A.10 and for a rigid probe patient in figure A.11. The resulting warping of the MRSI grid is clearly nonlinear in both cases. The percentage of strain energy due to external forces E averaged 8.6% for balloon probe cases and 10.0% for rigid probe cases. The low value for E demonstrates that, for both types of probes, most of the strain energy in the finite element simulation was due directly to the displacement of tissues caused by the probe rather than other uncertain external forces. Mean computation time for the image registration algorithm was comparable for both balloon and rigid probe patients on a 1.6 GHz Pentium-M laptop PC: 39.8 seconds with a standard deviation of 20.8 seconds for balloon probe cases and 34.2 seconds with a standard deviation of 11.8 seconds for rigid probe cases.

A.3.6 Discussion

Compensating for tissue deformations using biomechanical simulation with nonlinear parameter estimation results in better image registration than center-of-mass translation for all of the 10 cases tested. The DSC increased by an average

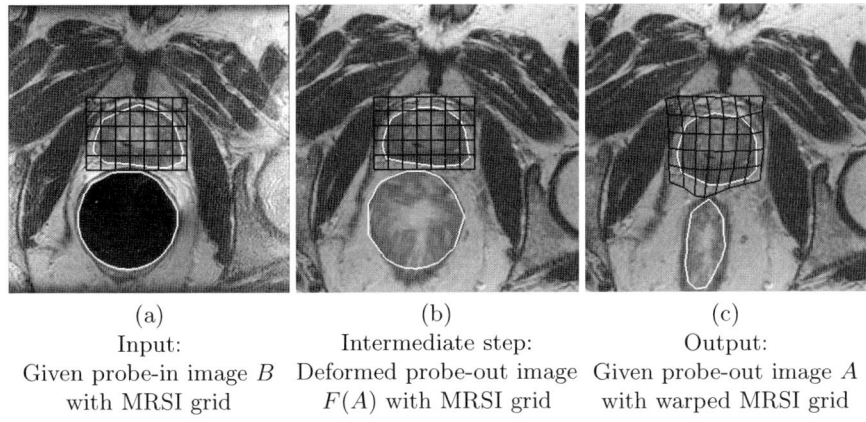

(a)   (b)   (c)
Input: Intermediate step: Output:
Given probe-in image B Deformed probe-out image Given probe-out image A
with MRSI grid $F(A)$ with MRSI grid   with warped MRSI grid

Fig. A.10. Sample balloon probe case. A comparison of input and output images shows the nonlinear warping of the MRSI grid. The probe-in image (a) closely matches the computed deformed probe-out image (b) outside the endorectal probe. The MRSI grid is warped to the undeformed probe-out image (c) for use during treatment planning.

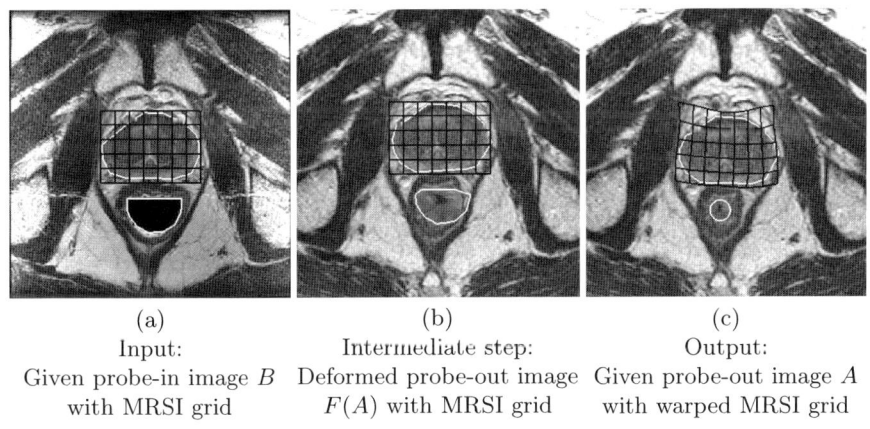

(a)   (b)   (c)
Input: Intermediate step: Output:
Given probe-in image B Deformed probe-out image Given probe-out image A
with MRSI grid $F(A)$ with MRSI grid   with warped MRSI grid

Fig. A.11. Sample rigid probe case

7.5% across all patients when using our method. These improvements come at a cost of computation time: our method required on average 37 seconds for each patient image slice in addition to manual image segmentation time for the probe-in and probe-out images.

Since our method only explicitly considers deformation in a 2-D (x, y) plane, it will not address out-of-plane deformations along the z-axis in a 3-D volume. However, past work has shown that z-axis deformations are small relative to the resolution of the volume images. Kim et al. found that the difference in the superior/inferior length of the prostate between probe-in and probe-out images was always less than the z-axis thickness of an axial MR image (3 mm) for 25 patient cases (15 rigid probe and 10 balloon probe) [122]. Crouch et al. measured seed displacements for 25 implanted seeds between balloon probe-in and probe-out images and found that the z-axis displacement averaged 2.67 mm, less than the 3 mm MR image slice thickness [59].

Our image registration method is sensitive to the segmentation of the image and the optimization algorithm may incorrectly add external forces or modify tissue stiffness properties if the segmentation is incorrect. Bhathara et al. quantified the error introduced by human segmentation: a human subject segmenting five 1.5 T MRI scans five times in random order achieved a mean DSC for segmentation reproducibility of 95% with a 95% confidence interval of (92%, 97%) while a second subject achieved a mean of 96% with confidence interval (95%, 97%) [35].

Using our method for image registration resulted in a greater improvement in mean DSC for balloon probe images (10.9%) than for rigid probe images (4.0%) when compared to registration by center-of-mass translation. Although balloon probes result in better quality images, these probes produce much larger deformations [122]. Kim et al. manually measured the anterior-posterior (AP) and right-left (RL) dimensions of the prostate in probe-in and probe-out images and found that the balloon probes on average compressed the prostate 3.5-fold more in the AP direction and stretched the prostate 2.5-fold more the RL direction than the rigid probe [122].

When compared to other image registration methods based on tissue deformation models, our method performs well. Our results visually appear to have smaller error than results from 2-D slices of 3-D volumes obtained by Wu et al., although a precise comparison is not possible because the accuracy of their method was not numerically quantified [215]. Wu et al. consider images taken with a balloon probe at different levels of inflation. This is different from comparing a probe-in image with a probe-out image in which the probe is removed entirely. Crouch et al. tested their finite element based method using an artificial tissue phantom with 25 radioactive seeds implanted inside [59, 60]. The phantom was deformed by a balloon probe resulting in average seed displacements of 9.377 mm, similar to the 9.22 mm average displacement of our test points in the interior of the prostate. To achieve 2.0 mm average point errors for the seeds, Crouch et al. required a mesh of 14,068 nodes and 14 hours of computation time for full 3-D deformations. Our method, which was tested on MR images of patient cases rather than tissue phantoms, achieved a less than 2 mm error for a representative point but required under 1 minute of computation time per image slice. Our DSC of 97.5% is higher than the 94% obtained by Bharatha et al. with a 3-D biomechanical finite element model for the registration of balloon endorectal probe-in images to rectal obturator (smaller) probe images [35]. However, subjects in that study were scanned in two different positions, supine and lithotomy, at two different field strengths, 1.5 T and 0.5 T, and with two different rectal probes, an MR expandable endorectal probe and a rectal obturator, which may have compromised image registration quality.

A.4   Conclusion and Open Problems

Physically-based simulation with nonlinear estimation of uncertain tissue parameters can improve the quality of deformable image registration. This approach is particularly useful for registration problems in which the tissue deformations are large and due to a known physical source.

For the application of registering (probe-in) MRSI data with (probe-out) radiation treatment planning images, improvements are greater for balloon probes compared to rigid probes due to the larger tissue deformations that occur with balloon probes. The algorithm achieved a mean DSC quality of 97.5% for five balloon probe patients and 98.1% for five rigid probe patients. The improvement over center-of-mass rigid registration is statistically significant ($P < 0.05$). Our method reduced displacement error between homologous test points in the probe-in and probe-out images by a mean 74.8% to a mean error of 1.95 mm for balloon probe cases and by a mean 70.0% to a mean error of 0.97 mm for rigid probe cases. The method required on average 37 seconds of computation time on a 1.6 GHz Pentium-M laptop PC to estimate and compensate for tissue deformations and produce a nonlinear mapping between probe-in and probe-out images.

The implementation described in this appendix independently registers 2-D slices of tissue from a 3-D MRI volume. Extending these results to an analogous 3-D physically-based simulation and image registration method would enable the method to explicitly account for deformations and displacements that occur between imaging planes in 3-D volumes. This extension would require generating patient-specific 3-D conformal tetrahedral meshes with a controlled number of elements. Validating the 3-D image registration approach would require using a new imaging protocol with slices sufficiently thin to capture out-of-plane deformations. We discuss these challenges in chapter 8.

Index

afterloader   8, 92
anesthesia   4, 57, 113
animal study   111
animation   12, 18, 19
Armijo's rule   134

b-spline   132
balloon probe   136, 138, 142–147
bang-bang control   76, 78, 84, 85, 89
Bellman equation   68, 83
bevel-tip   5, 30, 45–47, 49–51, 55, 57, 60–62, 78
bicycle model   46
biomechanical model   9, 129, 132, 133, 137, 138, 140
biopsy   1, 4, 23, 27, 29, 30, 46, 52, 54, 55, 57, 61, 113, 132
bladder   6, 7, 59, 96, 97, 100–104, 131, 133, 137
bone   20, 33, 45, 48, 57, 59, 84, 133, 139, 140
boundary condition   20, 32, 33, 40, 48, 138
brachytherapy   4, 6–10, 27, 28, 36, 39, 40, 57, 59, 91–94, 104, 106, 107, 112, 113, 135
brain   113

cancer   2, 4, 6, 7, 9, 10, 30, 39, 40, 57, 59, 61, 91, 94–97, 99–102, 105–107, 112, 129, 135
cannula   61, 113
catheter   8, 91–93, 95, 100, 104
Cauchy strain   21, 23, 31

clinical criteria   2, 6, 9, 91, 94–97, 100, 105, 107
clinician   93
collision detection   33, 72, 111
computational complexity   71, 80, 84, 109
cone-beam CT   113
configuration space   59, 75–77, 79–83, 87, 89
conformality   6, 95, 129, 135
Conjugate gradient method   22
constrained optimization   53
continuum mechanics   12, 13, 15, 18–20
convergence   39, 54, 77, 87, 93, 94, 105, 134
convex   53, 59, 98, 106, 131, 135
correlation coefficient   137
CT   2, 7, 8, 61, 91, 92, 100, 104, 109, 113, 137
curvature   5, 10, 45, 49–51, 57, 59–64, 69, 71, 73, 77, 79, 84, 85

da Vinci Surgical System   1
deformable tissue   30, 51, 52, 138
deformation   1–4, 7–9, 11–21, 23, 24, 27, 29–33, 36–39, 42, 43, 45–48, 50–52, 54, 55, 61, 107–111, 113, 129–135, 137, 139, 140, 142, 144–147
deformed mesh   20, 33–36, 48
degenerate element   35, 109
Delaunay triangulation   32, 48, 133, 140
Dice similarity coefficient   141–144, 146, 147

digital information   1, 113
Dijkstra's shortest path algorithm   66
discretization error   32, 59, 61, 66, 70, 71, 80, 87
displacement field   13, 15, 16, 18–20
dominant intraprostatic lesion   93, 135
dose calculation   93–98, 100, 106, 137
dosimetric index   92, 94, 95, 100–106
Dubins car   57–61, 73, 76, 77, 84, 89
dwell position   8, 93, 95, 98, 100
dwell time   6, 8, 10, 91, 93–95, 97–100, 104–106
dynamic programming   9, 59, 60, 62, 67, 68, 83, 108

edge   49, 50, 66, 70, 76, 81, 84, 94
elasticity   23, 137, 138
elastography   32, 133, 134, 140
element   2, 3, 8, 9, 12, 16–21, 23, 30–36, 39, 40, 43, 45, 48–50, 54, 55, 61, 83, 93, 108, 109, 132–138, 140, 141, 144, 146, 147
Euler time integration   48
external beam radiation treatment   135
eye   113

fat   40
FDA   8
feedback   11, 23, 36, 40, 59, 62, 63, 79, 85, 113
fiducial   138
finite element method   3, 9, 12, 19–21, 23, 30–33, 39, 43, 45, 48, 55, 109, 111, 133, 134, 136, 138, 146
fixed image   130–135, 139
fluoroscopy   2, 57, 61
force   8, 10, 12–15, 17, 18, 20, 21, 23, 31–34, 36, 43, 45, 47–51, 55, 63, 84, 108–110, 132–135, 137, 143, 144, 146
friction   31–33, 36, 43, 45, 47, 49–51, 55, 63, 109, 111

Gauss-Seidel method   22, 134
GE   139
geometric algorithm   2, 9, 13, 15, 18–20, 32, 48, 60, 76, 79, 80, 94, 104, 110, 132
goal   2, 11, 29, 37, 52, 64, 67, 75–79, 82, 83, 86, 87, 91, 108, 130, 132, 139

golden section search   39, 41, 42
gradient descent   53, 54
graph   66, 68, 76, 81, 83
graphics   8, 12, 23, 36, 108
Green's strain   23

heterogeneous tissue   111
high-dose-rate   6, 8–10, 91–94, 98, 104, 106, 112, 135
Hooke's Law   15

image   1–4, 6–9, 23, 24, 28, 29, 32, 36, 37, 40, 43, 46, 55, 57, 59–61, 63, 72, 73, 84, 89, 91, 92, 95, 100, 106–109, 111, 113, 129–147
image-guided medical procedure   1–4, 8, 9, 23, 24, 29, 43, 57, 59, 60, 72, 89, 107, 108
imaging   1, 2, 7, 8, 10, 11, 18, 27, 29, 43–47, 55, 57, 58, 60–63, 70, 72–74, 84, 85, 92, 100, 106–108, 113, 129–132, 136–139, 141, 144–147
incompressible   133
injection   4, 27, 113
integration   2, 18, 23, 33, 48
intensity modulated radiation therapy   113
interpolation   20, 48, 134, 135
Intuitive Surgical   1
Inverse Planning by Simulated Annealing   6, 91, 93, 94, 104–106
Inverse Planning by Simulated Annealing   93, 95, 98–100, 106
isodose contours   104, 105

kd-tree   84, 86
kinematics   29, 47, 60

large deformation   21, 23, 132
linearly elastic   15, 17, 31, 33, 45, 48, 110, 133, 135
linear programming   6, 9, 10, 91, 93–95, 97–106
linear system   22, 133, 134
liver   21, 60, 113
low-dose-rate   7

magnetic resonance imaging   7, 8, 29, 45, 46, 55, 57, 61, 92, 109, 130, 132, 135–139, 145–147

magnetic resonance spectroscopy imaging
    1, 2, 106, 129, 130, 132, 135–139,
    141, 144, 145, 147
Marching Cubes algorithm   109
Markov Decision Process   5, 9, 59, 60,
    62, 72, 73, 76, 78, 79, 83, 89, 108
mass-spring system   17–19, 109, 111
mass lumping   22, 23, 111
medical device   23, 113
medical robotics   29, 107
mesh   9, 19–21, 23, 32–36, 39, 40, 43,
    48–51, 54, 109–111, 132–135, 138,
    140–142, 144, 146, 147
meshless method   109
mesh modification   32, 34, 43, 109
mesh refinement   32, 109
micro-robot   113
Minkowski sum   63, 77
mobile robot   59, 76–79, 89
molecular imaging   113
motion planning   1–6, 8, 9, 24, 27, 30,
    37, 38, 43, 45, 46, 52, 57–62, 64,
    66–68, 71, 73, 75, 77, 79, 80, 87–89,
    91, 92, 107, 108, 111, 113
motion uncertainty   5, 9, 10, 59, 60, 62,
    67, 70, 72–79, 85–89, 108, 113
muscle   18, 133
mutual information   134, 137

National Library of Medicine   7
nearest neighbor   82, 84
needle   2–10, 12, 18, 20, 21, 24, 27–43,
    45–55, 57–73, 75, 78–80, 84, 85, 88,
    89, 100, 107–109, 112, 113, 132
needle insertion   3, 4, 9, 10, 12, 18, 20,
    21, 24, 27–33, 36–40, 42, 43, 45–49,
    52, 55, 57, 59–63, 71–73, 78, 84,
    108, 109, 113
needle steering   5, 9, 10, 24, 45–47, 52,
    57, 59, 60, 62–64, 66, 68, 73, 78, 79,
    84, 88, 89, 108, 112, 113
nerve   45, 57, 84
Newmark method   21, 22, 33
Newton's method   53
nodal displacement   21, 134
node   18–21, 31–36, 39, 40, 43, 48–51,
    66, 81, 109, 133–135, 140–142, 144,
    146

nonconvex   54, 63
nonholonomic   5, 9, 45, 46, 57, 59, 60,
    72, 73, 77–79, 84, 89
nonlinear FEM   23, 111
nonlinear material properties   31, 32,
    110
Nucletron   8, 93, 100

objective function   5, 54, 60, 62, 79,
    92–95, 98–101, 104–106, 134, 138,
    142, 143
obstacle   1, 3, 5, 9, 10, 29, 45–47, 52–55,
    57–61, 63, 66, 68, 69, 71, 73, 75,
    77–80, 82–86, 88, 89, 107, 108
oct-tree   111
optical coherence tomography   113
optimization   2–6, 8, 9, 27, 29, 30, 37,
    38, 42, 43, 47, 53–55, 58–60, 62,
    64, 66, 69–71, 73–79, 83, 89, 91–95,
    97–100, 104–108, 112, 129, 130,
    132–135, 137, 138, 143, 146

Partially Observable Markov Decision
    Process   78
patient case   11, 91, 94, 95, 101–106,
    129, 142, 146
penalty method   53, 54
permanent prostate implant   7
physically-based simulation   2, 3, 8, 9,
    11, 30, 107, 108
physician   3, 4, 6–9, 11, 23, 24, 27, 29,
    31, 36, 37, 40, 42, 43, 91, 93, 96,
    100, 106–108, 113
pixel   130, 131, 134, 135, 141, 142
Poisson's ratio   17, 18, 48, 133
polygon   46, 52, 55, 59, 85
polyhedron   94
probabilistic roadmap   75–77, 79, 89
probability of success   58–60, 62, 67–70,
    73, 76, 77, 79, 83, 87, 88
probe   40, 46, 129, 136–147
prostate   4, 6–10, 23, 24, 27–29, 32, 33,
    36, 38–42, 46, 48, 55, 59, 61, 91, 92,
    94–97, 99–107, 113, 129, 130, 132,
    135–137, 139–143, 145, 146
prostate cancer   4, 7, 59, 96, 97, 102,
    105, 129
prostate central gland   138, 140, 142,
    143

prostate peripheral zone   138, 140, 143
protein   75

quality metric   134, 135
quasiconvex   39

radiation source   4, 6, 10, 91, 107, 108
radiation treatment   4, 91, 129, 130,
    135, 136, 147
Rayleigh damping   21
real-time   3, 8, 12, 18, 22, 23, 27, 29,
    30, 33, 43–45, 57, 58, 61, 62, 108,
    109, 112
rectum   6, 40, 46, 52, 59, 92, 96, 97,
    100–104, 138–141, 146
reference mesh   19, 20, 22, 32–36, 39
registration   3, 9, 10, 12, 24, 106,
    129–139, 141–144, 146, 147
retina   1, 113
rigid probe   46, 136, 138, 142–147
rigid transform   130, 131
roadmap   5, 9, 10, 75–89, 108
robot   1, 8, 31, 58, 59, 62, 75–84, 86–89,
    92, 112, 113
robotic surgical assistant   1, 3, 4, 29,
    61, 113
robust optimization   106
rotation   38, 45, 47–49, 52, 53, 55, 62,
    130, 137

sampling   5, 9, 10, 75–77, 79–81, 86, 87,
    89, 108
seed   4, 6–8, 27–30, 36, 37, 39–42, 57,
    59, 91, 92, 135, 146
segmentation   14, 32, 34, 40, 49, 110,
    133, 138–140, 145, 146
sensorless planning   27–30, 43
shortest path   9, 10, 59, 60, 62, 66, 67,
    69–71, 75, 77, 79, 83, 87, 88
signal-to-noise ratio   129, 136
simulated annealing   6, 9, 91, 93–95,
    100–106
simulation   2–5, 8, 9, 11–13, 15, 18,
    19, 21–25, 27, 29–34, 36–41, 43,
    45, 47–55, 59, 72, 78, 86, 87, 89,
    107–111, 129, 133–135, 140, 144,
    147
simulator   29, 37, 40, 51, 53, 54
slip   31, 36, 48, 50, 51, 58, 111

soft tissue   2–5, 8, 9, 11–13, 15, 17, 18,
    21, 23, 24, 27, 29–31, 33, 36, 38, 43,
    45–49, 51, 52, 55, 57–61, 63, 73, 78,
    107, 108, 110, 113, 129, 131–134,
    137, 141
sparse matrix   83
steepest descent   134
steerable needle   4, 5, 8–10, 45, 47, 54,
    55, 57, 59–61, 63, 64, 73, 78–80,
    107, 111
stiffness   15, 17, 18, 21, 31, 63, 70, 85,
    132–135, 137, 138, 140, 143, 146
stochastic motion roadmap   5, 9, 10,
    75–77, 79–82, 84–89, 108
stochastic roadmap simulation   78
strain   14–18, 21, 23, 31, 110, 134, 142,
    144
stress   14–18, 20, 21, 110
surgery simulation   11, 12, 18, 19, 21
surgical device   1–3, 18, 107, 108, 110,
    113

target   1–5, 9, 10, 12, 23, 27–30, 36–39,
    41, 42, 45, 46, 52–55, 57–60, 62, 63,
    66, 67, 73, 78, 91, 94, 95, 100, 108,
    129
testbed   111, 112
tetrahedron   19, 109, 133, 138, 147
texture-mapping   23, 36, 43, 135
time step   18, 21–23, 31, 33, 34, 38–40,
    43, 48, 49, 51
training   11, 29, 41–43
transform   1, 13, 64, 129–133, 139, 141
transition probability matrix   67, 82
translation   47, 66, 130, 131, 136, 137,
    142–144
triangle   19–21, 32, 34–36, 40, 43,
    48–51, 54, 133, 135, 140

ultrasound   2, 8, 24, 27–29, 36, 37, 40,
    43, 45, 57, 59, 92, 100, 109, 133, 140
uncertainty   1–3, 5, 9, 10, 44, 58–60,
    62, 64, 66, 67, 69, 70, 72–80, 84–89,
    106–108, 113
unconstrained optimization   53, 54
unicycle model   46, 60
unimodal   39
urethra   6, 7, 59, 92, 96, 97, 100–104

value iteration   68, 69, 73, 83, 84

Visible human   7
visualization   11, 12, 23, 24, 36, 40, 43,
    100, 113, 135
Voronoi cell   82

workspace   57–59, 63, 64, 66, 69–71, 73,
    76–80, 84, 85

world frame   20, 30, 32–34, 36, 39,
    48–50

Young's modulus   15, 17, 18, 33, 40, 48,
    133, 140

zero-winding rule   36, 86

Springer Tracts in Advanced Robotics

Edited by B. Siciliano, O. Khatib and F. Groen

Further volumes of this series can be found on our homepage: springer.com

Vol. 50: Alterovitz, R.; Goldberg K.
Motion Planning in Medicine: Optimization
and Simulation Algorithms for
Image-Guided Procedures
153 p. 2008 [978-3-540-69257-7]

Vol. 49: Ott, C.
Cartesian Impedance Control of Redundant
and Flexible-Joint Robots
190 p. 2008 [978-3-540-69253-9]

Vol. 48: Wolter, D.
Spatial Representation and
Reasoning for Robot
Mapping
185 p. 2008 [978-3-540-69011-5]

Vol. 47: Akella, S.; Amato, N.;
Huang, W.; Mishra, B.; (Eds.)
Algorithmic Foundation of Robotics VII
524 p. 2008 [978-3-540-68404-6]

Vol. 46: Bessière, P.; Laugier, C.;
Siegwart R. (Eds.)
Probabilistic Reasoning and Decision
Making in Sensory-Motor Systems
375 p. 2008 [978-3-540-79006-8]

Vol. 45: Bicchi, A.; Buss, M.;
Ernst, M.O.; Peer A. (Eds.)
The Sense of Touch and Its Rendering
281 p. 2008 [978-3-540-79034-1]

Vol. 44: Bruyninckx, H.; Přeučil, L.;
Kulich, M. (Eds.)
European Robotics Symposium 2008
356 p. 2008 [978-3-540-78315-2]

Vol. 43: Lamon, P.
3D-Position Tracking and Control
for All-Terrain Robots
105 p. 2008 [978-3-540-78286-5]

Vol. 42: Laugier, C.; Siegwart, R. (Eds.)
Field and Service Robotics
597 p. 2008 [978-3-540-75403-9]

Vol. 41: Milford, M.J.
Robot Navigation from Nature
194 p. 2008 [978-3-540-77519-5]

Vol. 40: Birglen, L.; Laliberté, T.; Gosselin, C.
Underactuated Robotic Hands
241 p. 2008 [978-3-540-77458-7]

Vol. 39: Khatib, O.; Kumar, V.; Rus, D. (Eds.)
Experimental Robotics
563 p. 2008 [978-3-540-77456-3]

Vol. 38: Jefferies, M.E.; Yeap, W.-K. (Eds.)
Robotics and Cognitive Approaches to
Spatial Mapping
328 p. 2008 [978-3-540-75386-5]

Vol. 37: Ollero, A.; Maza, I. (Eds.)
Multiple Heterogeneous Unmanned Aerial
Vehicles
233 p. 2007 [978-3-540-73957-9]

Vol. 36: Buehler, M.; Iagnemma, K.;
Singh, S. (Eds.)
The 2005 DARPA Grand Challenge – The Great
Robot Race
520 p. 2007 [978-3-540-73428-4]

Vol. 35: Laugier, C.; Chatila, R. (Eds.)
Autonomous Navigation in Dynamic
Environments
169 p. 2007 [978-3-540-73421-5]

Vol. 34: Wisse, M.; van der Linde, R.Q.
Delft Pneumatic Bipeds
136 p. 2007 [978-3-540-72807-8]

Vol. 33: Kong, X.; Gosselin, C.
Type Synthesis of Parallel
Mechanisms
272 p. 2007 [978-3-540-71989-2]

Vol. 32: Milutinović, D.; Lima, P.
Cells and Robots – Modeling and Control of
Large-Size Agent Populations
130 p. 2007 [978-3-540-71981-6]

Vol. 31: Ferre, M.; Buss, M.; Aracil, R.;
Melchiorri, C.; Balaguer C. (Eds.)
Advances in Telerobotics
500 p. 2007 [978-3-540-71363-0]

Vol. 30: Brugali, D. (Ed.)
Software Engineering for Experimental Robotics
490 p. 2007 [978-3-540-68949-2]

Vol. 29: Secchi, C.; Stramigioli, S.; Fantuzzi, C.
Control of Interactive Robotic Interfaces – A
Port-Hamiltonian Approach
225 p. 2007 [978-3-540-49712-7]

Vol. 28: Thrun, S.; Brooks, R.; Durrant-Whyte, H. (Eds.)
Robotics Research – Results of the 12th International Symposium ISRR
602 p. 2007 [978-3-540-48110-2]

Vol. 27: Montemerlo, M.; Thrun, S.
FastSLAM – A Scalable Method for the Simultaneous Localization and Mapping Problem in Robotics
120 p. 2007 [978-3-540-46399-3]

Vol. 26: Taylor, G.; Kleeman, L.
Visual Perception and Robotic Manipulation – 3D Object Recognition, Tracking and Hand-Eye Coordination
218 p. 2007 [978-3-540-33454-5]

Vol. 25: Corke, P.; Sukkarieh, S. (Eds.)
Field and Service Robotics – Results of the 5th International Conference
580 p. 2006 [978-3-540-33452-1]

Vol. 24: Yuta, S.; Asama, H.; Thrun, S.; Prassler, E.; Tsubouchi, T. (Eds.)
Field and Service Robotics – Recent Advances in Research and Applications
550 p. 2006 [978-3-540-32801-8]

Vol. 23: Andrade-Cetto, J,; Sanfeliu, A.
Environment Learning for Indoor Mobile Robots – A Stochastic State Estimation Approach to Simultaneous Localization and Map Building
130 p. 2006 [978-3-540-32795-0]

Vol. 22: Christensen, H.I. (Ed.)
European Robotics Symposium 2006
209 p. 2006 [978-3-540-32688-5]

Vol. 21: Ang Jr., H.; Khatib, O. (Eds.)
Experimental Robotics IX – The 9th International Symposium on Experimental Robotics
618 p. 2006 [978-3-540-28816-9]

Vol. 20: Xu, Y.; Ou, Y.
Control of Single Wheel Robots
188 p. 2005 [978-3-540-28184-9]

Vol. 19: Lefebvre, T.; Bruyninckx, H.; De Schutter, J. Nonlinear Kalman Filtering for Force-Controlled Robot Tasks
280 p. 2005 [978-3-540-28023-1]

Vol. 18: Barbagli, F.; Prattichizzo, D.; Salisbury, K. (Eds.)
Multi-point Interaction with Real and Virtual Objects
281 p. 2005 [978-3-540-26036-3]

Vol. 17: Erdmann, M.; Hsu, D.; Overmars, M.; van der Stappen, F.A (Eds.)
Algorithmic Foundations of Robotics VI
472 p. 2005 [978-3-540-25728-8]

Vol. 16: Cuesta, F.; Ollero, A.
Intelligent Mobile Robot Navigation
224 p. 2005 [978-3-540-23956-7]

Vol. 15: Dario, P.; Chatila R. (Eds.)
Robotics Research – The Eleventh International Symposium
595 p. 2005 [978-3-540-23214-8]

Vol. 14: Prassler, E.; Lawitzky, G.; Stopp, A.; Grunwald, G.; Hägele, M.; Dillmann, R.; Iossifidis. I. (Eds.)
Advances in Human-Robot Interaction
414 p. 2005 [978-3-540-23211-7]

Vol. 13: Chung, W.
Nonholonomic Manipulators
115 p. 2004 [978-3-540-22108-1]

Vol. 12: Iagnemma K.; Dubowsky, S.
Mobile Robots in Rough Terrain – Estimation, Motion Planning, and Control with Application to Planetary Rovers
123 p. 2004 [978-3-540-21968-2]

Vol. 11: Kim, J.-H.; Kim, D.-H.; Kim, Y.-J.; Seow, K.-T.
Soccer Robotics
353 p. 2004 [978-3-540-21859-3]

Vol. 10: Siciliano, B.; De Luca, A.; Melchiorri, C.; Casalino, G. (Eds.)
Advances in Control of Articulated and Mobile Robots
259 p. 2004 [978-3-540-20783-2]

Vol. 9: Yamane, K.
Simulating and Generating Motions of Human Figures
176 p. 2004 [978-3-540-20317-9]

Vol. 8: Baeten, J.; De Schutter, J.
Integrated Visual Servoing and Force Control – The Task Frame Approach
198 p. 2004 [978-3-540-40475-0]

Vol. 7: Boissonnat, J.-D.; Burdick, J.; Goldberg, K.; Hutchinson, S. (Eds.)
Algorithmic Foundations of Robotics V
577 p. 2004 [978-3-540-40476-7]

Vol. 6: Jarvis, R.A.; Zelinsky, A. (Eds.)
Robotics Research – The Tenth International Symposium
580 p. 2003 [978-3-540-00550-6]